AF559874

CHEMISTRY
AN ENVIRONMENTAL PROSPECTIVE

CHEMISTRY
AN ENVIRONMENTAL PROSPECTIVE

By

P.B. Saxena

DISCOVERY PUBLISHING HOUSE PVT. LTD.
NEW DELHI-110 002

Published by:
Namit Wasan

DISCOVERY PUBLISHING HOUSE PVT. LTD.
4383/4B, Ansari Road, Darya Ganj
New Delhi-110 002 (India)
Phone : +91-11-23279245; 23253475; 43596065
E-mail : discoverybooksindia@gmail.com
discoverypublishinghouse@gmail.com
namitwasan9@gmail.com
web : www.discoverypublishinggroup.com

First Published: **2009**

Reprinted: **2021**

ISBN: 978-81-8356-442-7

Chemistry: ***An Environmental Prospective***

Printed at:
Infinity Imaging Systems
Delhi

Preface

The present title "Chemistry: *An Environmental Prospective*" is the systematic study of the natural and man made world. It has emerged as a major disciplines in recent years, reflecting the growing concern about the impact of human activity on our surrounding. This text provides a clear and authoritative introduction to the fundamental concepts and vocabulary necessary to explore complex environmental issues and phenomenon. It progresses from an extensive study of the undisturbed natural world to a thorough examination of the impact of human activity upon it. It also covers the basic scientific concept necessary for understanding environmental issues, enabling students with differing academic backgrounds to gain immediate benefit from the text. It is a core undergraduate text for students of environmental sciences and related disciplines including environmental biology and environmental chemistry. It is also valuable as supplementary material for post-graduates envolved in pollution control and environmental management.

To make the work more comprehensive and informative, the author has consulted many authoritative books, research journals, abstracts, monographs etc., so there can be no claim to originality except in the manner of treatment.

The author expresses his thanks to his friends and colleagues whose continue inspirations have initiated him to bring out this book.

The author expresses his gratitude to Mr. Wasan and staff of M/s Discovery Publishing House Pvt. Ltd. for their whole hearted cooperation in the publication of this book.

In the mean time, the author will remain sincerely responsible for any shortcomings of the book and be grateful to the readers for their suggestions and constructive criticism for the continuous betterment of the book. He takes this opportunity to appeal to the readers to send their suggestions straightaway to his Publisher.

Author

CONTENTS

1

Introduction

Environmental science in its broadcast sense is the science of the complex interactions, that occur among the terrestrial, atmospheric, aquatic, and living environments. It includes all the disciplines, such as chemistry, biology, ecology, sociology, and government, that affect or describe these interactions. For the purpose of this book, environmental science will be defined as the study of the earth, air, water, and living environments and the effects of technology thereon. To a significant degree, environmental science has evolved from investigations of the ways by which, and places in which living organisms carry out their life cycles. This is the discipline of *natural history*, which in recent times has evolved into *ecology*, the study of environmental factors that affect organisms and of how organisms interact with these factors and with each other.

For better or for worse, the environment in which all humans must live has been affected irreversibly by technology. Therefore, technology is considered strongly in this book in terms of how it affects the environment and the ways by which, applied intelligently by those knowledgeable of environmental science, it can serve, rather than damage, this Earth upon which all living beings depend for their welfare and existence. The physical environments—air, water, and earth—are tied closely with living systems, including humans. These four environmental spheres have strong mutual interactions with technology.

A key aspect of environmental science is the "interrelatedness" of things, the influence that one thing, action, or change may have an another. One of the ways in which this interrelatedness is most strongly expressed is through feedback in environmental systems. *Positive*

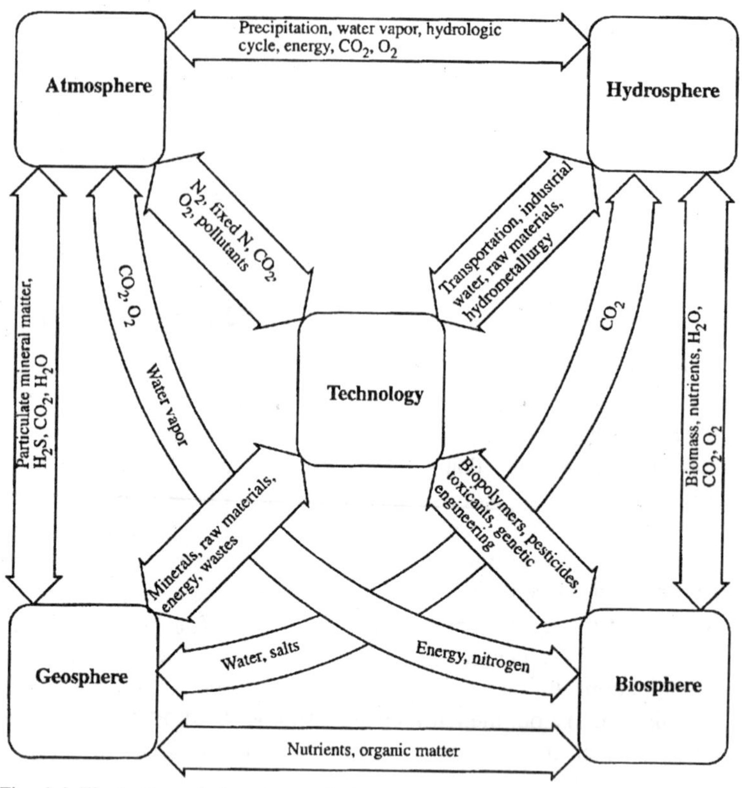

Fig. 1.1. Illustration of the close relationships among the air, water, and earth environments with each other and with living systems, as well as the tie-in with technology.

feedback occurs when a change or action tends to amplify itself. An example of detrimental positive feedback is provided by soil erosion set off, for example, by plowing virgin prairie land in hilly terrain. Topsoil is removed by running water causing gullies to form. As these form they can induce even larger erosive features. As topsoil is lost, it is harder to grow vegetative cover, making the soil even more prone to erosion. *Negative feedback* occurs when a system automatically adjusts to minimize an influence. Left to their own devices, rivers and their floodplains tend to develop in ways that compensate for increased river flow and minimize environmental harm from flooding.

Key Concepts and Definitions

Three are numerous key concepts and definitions that are used in discussing the environmental sciences.

Environmental science may be divided among the study of the atmosphere, the hydrosphere, the geosphere, and the biosphere. The *atmosphere* is the thin layer of gases that covers Earth's surface. In addition to its role as a reservoir of gases, the atmosphere moderates Earth's temperature, absorbs energy and damaging ultraviolet radiation from the sun, transports energy away from equatorial regions, serves as a pathway for vapor-phase movement of water in the hydrologic cycle. The *hydrosphere* contains Earth's water. Over 97% of Earth's water is in oceans, and most of the remaining fresh water is in the form of ice. Therefore, only a relatively small percentage of the total water on Earth is actually involved with terrestrial, atmospheric, and biological processes. Exclusive of seawater, the water that circulates through environmental processes and cycles occurs in the atmosphere, underground as groundwater, and as surface water in streams, rivers, lakes, ponds, and reservoirs.

The *geosphere* consists of the solid earth, including soil, which supports most plant life. The part of the geosphere that is directly involved with environmental processes through contact with the atmosphere, the hydrosphere, and living things is the solid *lithosphere*, varying from 50 to 100 km in thickness, and particularly the outer skin of the lithosphere composed largely of lighter silicate-based minerals and called the *crust*. All living entities on Earth compose the *biosphere*. Living organisms and the aspects of the environment pertaining directly to them are called *biotic*, and other portions of the environment are *abiotic*.

There are strong interactions among living organisms and the various spheres of the abiotic environment. To a large extent these are best described by cycles of matter that involve biological, chemical, and geological processes and phenomena. Such cycles are called *biogeochemical cycles*.

There is not space at this point to define all the terms and describe all the concepts essential to the understanding of environmental science. Several of these that should be noted here are the following:

1. *Matter*. Substance, that which occupies space and has mass.
2. *Chemistry*. The science of matter—the study of the composition, structure, and properties of mater and the changes that matter undergoes.
3. *Energy*. The capacity to do work, such as by causing a body of matter to move. The rate at which energy is transferred or moved, that is, energy per unit time is called *power*.

4. *Resources*. Matter of specific kinds and energy needed by humans for their well-being or existence. *Renewable resources* are those that are replenished naturally within a reasonable time span.
5. *Climate*. The overall, long-term characteristics of weather, including temperature, precipitation, storms, and wind patterns in an area.
6. *Pollutant*: A substance present in greater than natural concentration as a result of human activity and having a net detrimental effect upon its environment or upon something of value in that environment. Every pollutant originates from a *source*. A *receptor* is anything that is affected by a pollutant. A *sink* is a long-time repository of a pollutant.
7. *Biological community*: The total of all living organisms inhabiting a specified area. A biological community and the environmental conditions that characterize it are termed a *biome*. A group of organisms of the same species in a biological community is called a *population*. The role played by a biological population in its surroundings, including the pattern by which it uses available resources, is called its *ecological niche*.
8. *Productivity*. Rate of production of biomass per unit time per unit area by organisms called *producers*, which use energy (usually from photosynthesis) to produce biological matter from inorganic substances.
8. *Limiting factors*. Environmental factors, such as nutrients, that limit the abundance, growth, or distribution of an organism, or that determine whether it even exists in a particular environment.
9. *Ecology*. The study of the interactions of organisms with their environment and with each other.

Environmental Science, Technology, and Society

Modern societies and economic systems are driven largely by material wants and needs. The more industrially and economically advanced societies, such as those in Western Europe, the United States, and Japan, are characterized by a high level of consumption and the accompanying quest for money and the spending of it for material goods. Countries with developing economies are similarly driven, aspiring to the same standards of living as are enjoyed by citizens of richer nations. Residents of the poorest nations, such as a number of countries in Africa, Latin America, and Parts of Asia, are driven by a more fundamental need—the acquisition of enough food, water, fuel, and other basic necessities to ensure their survival.

Given these conditions and human attitudes, what is the role of environmental science in modern societies, and how is it related to technology and human needs and aspirations? Does environmental science even have a place in the richer societies that can afford the necessities of life and many luxuries as well? Should environmental factors even be considered in subsistence societies in which getting the most basic necessities for human survival is the overwhelming concern of most of the population? Can environmental science coexist with modern technology to the benefit of humankind? The answer to all of these questions is a resounding, "yes!"

Consider first the modern industrialized societies. The material benefits enjoyed by people in these societies cannot be sustained for long without coming to terms with the environment upon which the resources, energy, and quality surroundings enjoyed by the people depend. Depletion of resources of scarce metals and strategic minerals will force prices upward so that the possessions required for the so-called "Good life"—luxury automobiles, modern single-family dwellings, and other amenities—can no longer be purchased. Increased income levels enable people to move farther from population centers to enjoy less urbanized surroundings. The same conditions that draw people to the more rural areas deteriorate because of the presence of more people. These factors in turn exacerbate social problems and tensions. For example, there is tendency for the upper economic strata of society to user their influence to ensure conditions under which their material benefits will be sustained or enhanced, even if this means increased poverty in the lower economic strata and a diminished middle class. A large fraction of the poorer population "tunes out" and becomes involved with self-destructive behavior that makes their situation even worse. The environment in which people find themselves certainly influences their behavior and well-being.

The question is commonly asked whether or not developing and poor, undeveloped societies can afford to consider environment. The answer is that they cannot afford to do otherwise. In its broadest sense a good environment is one that is favorable for the existence of life in a flourishing, steady-state condition. Ruthless exploitation of the environment—farming of marginal land without proper conservation practices, poorly considered damming of rivers to provide hydroelectric power, and intense exploitation of mineral and energy resources—will result in environmental deterioration that is eventually counterproductive to the material standard of living that such measures are designed to enhance.

Whereas ill-considered exploitation of the environment is definitely counter-productive, the opposite approach of "going back to nature" and shunning all technological development is wildly impractical. It simply is not going to happen. Therefore, it is essential to realize that technology will be employed to meet human needs, that these needs can only be met on the basis of a healthy environment suitable for supporting humankind, and that societies must learn to live in harmony with the environment and with technology.

Water

Water, with a deceptively simple chemical formula of H_2O, is a vitally important substance in all parts of the environment. Water covers about 70% of Earth's surface. It occurs in all spheres of the environment—in the oceans as a vast reservoir of saltwater, on land as surface water in lakes and rivers, underground as groundwater, in the atmosphere as water vapor, and in the polar icecaps as solid ice. Water is an essential part of all living systems and is the medium from which life evolved and in which life exists.

Energy and matter are carried through various spheres of the environment by water. Water leaches soluble constituents from mineral mater and carries them to the ocean or leaves them as mineral deposits some distance from their sources. Water carries plant nutrients from soil into the bodies of plants by way of plant roots. Solar energy absorbed in the evaporation of ocean water is carried as latent heat and released inland. The accompanying release of latent heat provides the energy that powers massive storms.

The quantity of water in various forms on or beneath Earth's surface and in the atmosphere is enormous, amounting to about 1.4 billion cubic kilometers of liquid water. By far the largest portion of this water, about 97.6%, occurs in the oceans. The next largest amount, about 2%, is held as ice and snow, largely in the polar and Greenland icecaps. Groundwater held in underground aquifers amounts to 0.28% (to a depth of 1 km). The rest of Earth's water, only about 0.25% of the total, is found in decreasing order of abundance in saline and freshwater lakes and reservoirs; as soil moisture; as water held in living organisms, as vapors, droplets, and miniscule ice crystals in the atmosphere; in swamps and marshes; and in rivers and streams.

Air and Atmosphere

The atmosphere is a protective blanket that nurtures life on the Earth and protects it from the hostile environment of outer space. The atmosphere is the source of carbon dioxide for plant photosynthesis and of oxygen for respiration. It provides the nitrogen that nitrogen-fixing bacteria and ammonia-manufacturing industrial plants use to produce chemically bound nitrogen, an essential component of life molecules. As a basic part of the hydrologic cycle the atmosphere transports water from the oceans to land, thus acting as the condenser in a vast solar-powered still. The atmosphere serves a vital protective function. It absorbs most of the cosmic rays from outer space and protects organisms from their effects. It also absorbs most of the electromagnetic radiation from the sun, allowing transmission of significant amounts of radiation only in the near-ultraviolet, visible, and near-infrared radiation regions. The atmosphere filters out damaging ultraviolet radiation that would otherwise be very harmful to living organisms. Furthermore, because it reabsorbs much of the infrared radiation by which absorbed solar energy is re-emitted to space, the atmosphere stabilizes the Earth's temperature, preventing the great temperature extremes of planes and moons lacking substantial atmosphere.

Atmospheric science deals with the movement of air masses in the atmosphere, atmospheric heat balance, and atmospheric chemical composition and reactions. *Meterology* is the study of the movement of air masses as well as physical forces in the atmosphere such as heat, wind, and transitions of water, primarily liquid to vapor, or *vice versa*. Short-term variations in the state of the atmosphere, including temperature, clouds, winds, humidity, horizontal visibility, type and quantity of precipitation, and atmospheric pressure, are described as *weather*. Long-term conditions of weather referred to regionally or even globally are classified as *climate*. The nature of climate and potential human effects on it are vitally important and environmental protection.

EARTH

The *geosphere*, or solid Earth, is that part of the Earth upon which humans live and from which they extract most of their food, minerals, and fuels. The earth is divided into layers, including the solid iron-rich inner core, molten outer core, mantle, and crust. Environmental science is most concerned with the *lithosphere*, which consists of the outer mantle and the crust. The latter in the earth's outer skin that is accessible to humans. It is extremely thin compared to the diameter of the earth, ranging from 5 to 40 km thick.

Geology is the science of the geosphere. As such it pertains mostly to the solid mineral portions of Earth's crust. But it must also consider water, which is involved in weathering rocks and in producing mineral formations; the atmosphere and climate, which have profound effects on the geosphere and interchange matter and energy with it; and living systems, which largely exist on the geosphere and in turn have significant effects on it. Geological science uses chemistry to explain the nature and behavior of geological materials, physics to explain their mechanical behavior, and biology to explain the mutual interactions between the geosphere and the biosphere. Modern technology, for example, the ability to move massive quantities of dirt and rock around, has a profound influence on the geosphere.

Most of the solid Earth crust consists of rocks. Rocks are composed of minerals, where a *mineral* is a naturally occurring inorganic solid with a definite internal structure and chemical composition. A *rock* is a mass of pure mineral or an aggregate of two or more minerals. The most important part of the geosphere for life on earth is *soil* formed by the disintegrative weathering action of physical, geochemical, and biological process on rock. It is the medium upon which plants grow, and virtually all terrestrial organisms depend upon it for their existence. The productivity of soil is strongly affected by environmental conditions and pollutants.

LIFE

Biology is the science of life. It is based on biologically synthesized chemical species that fall into the categories of proteins, lipids (fats, oils, some hormones), carbohydrates, and nucleic acids that largely constitute living organisms and enable them to function. Many of these substances exist as large molecules, called *macromolecules*. As living beings, the ultimate concern of humans with their environment is the interaction of the environment with life. Therefore, biological science is a key component of environmental science.

The focal point of biochemistry and biochemical aspects of toxicants is the *cell*, the basic building block of living systems where most life processes are carried out. Individual cells are so small that they cannot be seen without a microscope. Bacteria, yeasts, and some algae consist of single cells. However, most living things are made up of many cells. In a more complicated organisms the cells have different functions. Liver cells, muscle cells, brain cells, and skin cells in the human body are quite different from each other and do different things.

Consideration of biology from the perspective of cells and macromolecules is done at a microscopic level. At the other end of the scale, environmental scientists may regard the biosphere from the standpoint of *populations* of species interacting with one another in a *biological community*. As noted above, ecology is the study of environmental factors that affect organisms and of how organisms interact with these factors and with each other.

ECOLOGY

Ecology is the science that deals with the relationships between living organisms with their physical environment and with each other. Ecology can be approached from the viewpoints of (1) the environment and the demands it places on organisms in it or (2) organisms and how they adapt to their environmental conditions. An *ecosystem* consists of an assembly of mutually interacting organisms ard their environment in which materials are interchanged in a largely cyclical manner. An ecosystem has physical, chemical, and biological components along with energy sources and pathways of energy and materials interchange. The environment in which a particular organism lives is called its *habitat*. The role of an organism is a habitat is called its *niche*.

For the study of ecology it is often convenient to divide the environment into four broad categories. The *terrestrial environment* is based on land and consists of *biomes*, such as grasslands, one of several kinds of forests, savannas, or deserts. The *freshwater environment* can be further subdivided between *standing-water habitats* (lakes, reservoirs) and *running-water habitats* (streams, rivers). The oceanic *marine environment* is characterized by saltwater and may be divided broadly into the shallow waters of the continental shelf composing the *neritic zone* and the deeper waters of the ocean that constitute the *oceanic region*. An environment in which two or more kinds of organisms exist together to their mutual benefit is termed a *symbiotic environment*.

A particularly important factor in describing ecosystems is that of *populations* consisting of numbers of a specific species occupying a specific habitat. Populations may be stable, or they may grow exponentially as a *population explosion*. A population explosion that is unchecked results in resource depletion, waste accumulation, and predation culminating in an abrupt decline called a *population crash*. *Behavior* in areas such as hierarchies, territoriality, social stress, and feeding patterns plays a strong role in determining the fates of populations.

Two major subdivisions of modern ecology are *ecosystem ecology*, which views ecosystems as large units, and *population ecology*, which attempts to explain ecosystem behavior from the properties of individual units. In practice, the two approaches are usually merged. *Descriptive ecology* describes the types and nature of organisms and their environment, emphasizing structures of ecosystem sand communities and dispersions and structures of populations. *Functional ecology* explains how things work in an ecosystem, including how populations respond to environmental alteration and how matter and energy move through ecosystems. An understanding of ecology is essential in the management of modern industrialized societies in ways that are compatible with environmental preservation and enhancement. The branch of ecology that deals with predicting the impacts of technology and development and making recommendations such that these activities will have minimum adverse impacts, or even positive impacts, on ecosystems may be termed *applied ecology*.

Matter and Cycles of Matter

Biogeochemical cycles describe the circulation of matter, particularly plant and animal nutrients, through ecosystems. These cycles are ultimately powered by solar energy, fine-tuned and directed by energy expensed by organisms. In a sense, the solar-energy-powered hydrologic cycle acts as an endless conveyer belt to move materials essential for life through ecosystems.

Most biogeochemical cycles can be descried as elemental cycles involving nutrient elements such as carbon, oxygen, nitrogen, sulfur and phosphorus. Many are *gaseous cycles* in which the element in question spends part of cycle in the atmosphere—O_2 for oxygen, N_2 for nitrogen, CO_2 for carbon. Others, notably the phosphorus cycle, do not have a gaseous component and are called *sedimentary cycles*. All sedimentary cycles involve *salt solutions* or *soils solutions* that contain dissolved substances leached from weathered minerals, that may be deposited as mineral formations, or they may be taken up by organisms as nutrients. The sulfur cycle, which may have H_2S or SO_2 in the gaseous phase or minerals ($CaSO_4 \cdot 2H_2O$) in the solid phase, is a combination of gaseous and sedimentary cycles.

Carbon, the basic building block of life molecules, is circulated through the *carbon cycle*. This cycle shows that carbon may be present as gaseous atmospheric CO_2, dissolved in groundwater as HCO_3^- or molecular $CO_2(aq)$, in underlying rock strata as limestone ($CaCO_3$), and as organic matter, represented in a simplified manner as $\{CH_2O\}$. Photosynthesis fixes inorganic carbon as biological carbon, which is a constituent of all life molecules.

An important aspect of the carbon cycle is that it is the cycle by which energy is transferred to biological systems. Organic, or biological

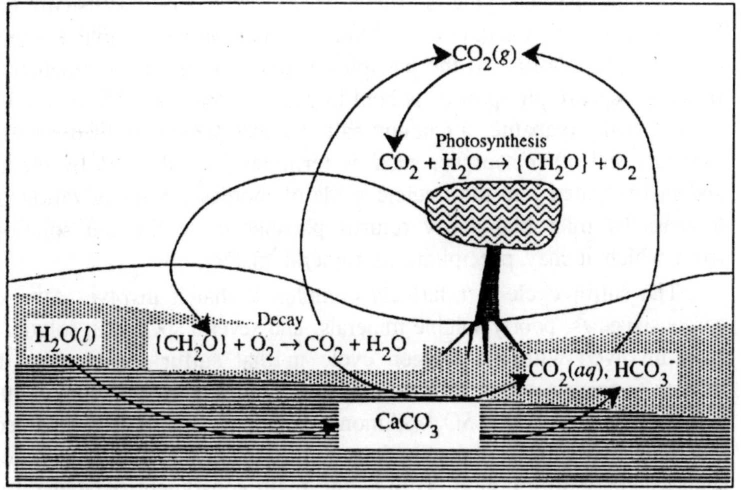

Fig. 1.2. The carbon cycle.

carbon, $\{CH_2O\}$, is an energy-rich molecule that can react biochemically with molecular oxygen, O_2, to regenerate carbon dioxide and produce energy. This can occur in an organism as shown by the "decay" reaction, or it may take place as combustion, such as when wood is burned.

The *oxygen cycle* involves the interchange of oxygen between the elemental form of gaseous O_2 in the atmosphere and chemically bound O in CO_2, H_2O, and organic matter. Elemental oxygen becomes chemically bound by various energy-yielding processes, particularly combustion and metabolic processes in organisms. It is released in photosynthesis.

Nitrogen, though constituting much less of biomass than carbon or oxygen, is an essential constituent of proteins. The atmosphere is 78% by volume elemental nitrogen, N_2, and constitutes an inexhaustible reservoir of this essential element. The N_2 molecule is very stable so that breaking it down to atoms that can be incorporated in inorganic and organic chemical forms of nitrogen is the limiting step in the *nitrogen cycle*. This does occur by highly energetic processes in lighting discharges such that nitrogen becomes chemically combined with hydrogen or oxygen as ammonia or nitrogen oxides. Elemental nitrogen is also incorporated into chemically bound forms or *fixed* by biochemical processes mediated by microorganisms. The biological nitrogen is returned to the inorganic form during the decay of biomass by a process called *mineralization*.

The phosphorus cycle is crucial because phosphorus is usually the limiting nutrient in ecosystems. There are no common stable gaseous forms of phosphorus, so the phosphorus cycle is strictly sedimentary. In the geosphere phosphorus is held largely in poorly soluble minerals, such as hydroxyapatite, a calcium salt. Soluble phosphorus from these minerals and other sources, such as fertilizers, is taken up by plants and incorporated into the nucleic acids of biomass. Mineralization of biomass by microbial decay returns phosphorus to the salt solution from which it may precipitate as mineral mater.

The sulfur cycle is relatively complex is that it involves several gaseous species, poorly soluble minerals, and several species in solution. It is involved with the oxygen cycle in that sulfur combines with oxygen to form gaseous sulfur dioxide, SO_2, an atmospheric pollutant, and soluble sulfate ion, SO_4^{2-}. Among the significant species involved in sulfur cycle are gaseous hydrogen sulfide, H_2S; mineral sulfide, such as Pbs: sulfuric acid, H_2SO_4, the main constituent of acid rain; and biologically bound sulfur in sulfur-containing proteins.

Energy and Cycles of Energy

Biogeochemical cycles and virtually all other processes on Earth are driven by energy from the sun. The sun acts as a so-called blackbody radiator with an effective surface temperature of 5780 K (Celsius degrees above absolute zero). It transmits energy to Earth as electromagnetic radiation. The maximum energy flux of the incoming solar energy is at a wavelength of about 500 nanometers, which is in the visible region of the spectrum. A 1 square meter area perpendicular to the line of solar flux at the top of the atmosphere receives energy at a rate of 1,340 watts, sufficient, for example, to power an electric iron. This is called the *solar flux*.

Energy in natural systems is transferred by *heat*, which is the form of energy that flows between two bodies as a result of their difference in temperature, or by *work*, which is transfer of energy that does not depend upon a temperature difference, as governed by the laws of *thermodynamics*. The *first law of thermodynamics* states that, although energy may be transferred or transformed, it is conserved and is not lost. Chemical energy in the food ingested by organisms is converted by *metabolic processes* to work or heat that can be utilized by the organisms, but there is not net gain or loss of energy overall. The *second law of thermodynamics* describes the tendency toward disorder in natural systems. It demonstrates that each time energy is transformed, some is lost in the sense that it cannot be utilized for work, so only a fraction of the energy that organisms derive from metabolizing food can be converted to work; the rest is dissipated as heat.

Light and Electromagnetic Radiation

Electromagnetic radiation, particularly light, is of utmost importance in considering energy in environmental systems. Therefore, the following important points related to electromagnetic radiation should be noted:

1. Energy can be carried through space at the speed of light, 3.00×10^8 meters per second (m/s) in a vacuum, by *electromagnetic radiation*, which includes visible light, ultraviolet radiation, infrared radiation, microwaves, and radio waves.
2. Electromagnetic radiation has a *wave character*. The waves move at the speed of light, c, and have characteristics of *wavelength* (λ), amplitude, and *frequency* (ν, Greek "nu") as illustrated below:

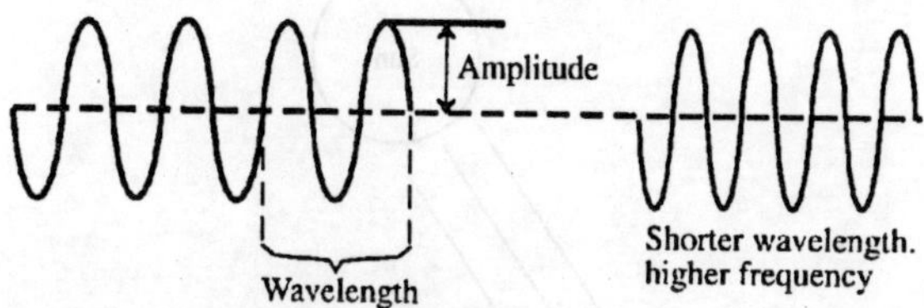

3. The wavelength is the distance required for one complete cycle and the frequency is the number of cycles per unit time. They are related by the following equation:

$$\nu\lambda = c$$

where ν is in units of cycles per second (s^{-1}, a unit called the *hertz*, Hz and λ is in meters (m).
4. In addition to behaving as a wave, electromagnetic radiation also has characteristics of particles.
5. The dual wave/particle nature of electromagnetic radiation is the basis of the *quantum theory* of electromagnetic radiation, which states that radiant energy may be absorbed or emitted only in discrete packets called *quanta* or *photons*. The energy, E, of each photon is given by

$$E = h\nu$$

where h is Planck's constant, 6.63×10^{-34} J-s (joule $\times$ second).
6. From the preceding, it is seen that the energy of a photon is higher when the frequency of the associated wave is higher (and the wavelength shorter).

Energy Flow and Photosynthesis

Whereas materials are recycled through ecosystems, the flow of useful energy may be viewed as essentially a one-way process. Incoming solar energy can be regarded as high-grade energy because it can cause useful reactions to occur, the most important of which in living systems is photosynthesis. Solar energy captured by green plants energizes chlorophyll, which in turn powers metabolic processes that produce carbohydrates from water and carbon dioxide. These carbohydrates represent stored chemical energy that can be converted to heat and work by metabolic reactions with oxygen in organisms. Ultimately, most of the energy is converted to low-grade heat, which is eventually re-radiated away from Earth by infrared radiation.

Energy Utilization

During the last two centuries the human impact on energy utilization and conversion has been enormous and has resulted in many

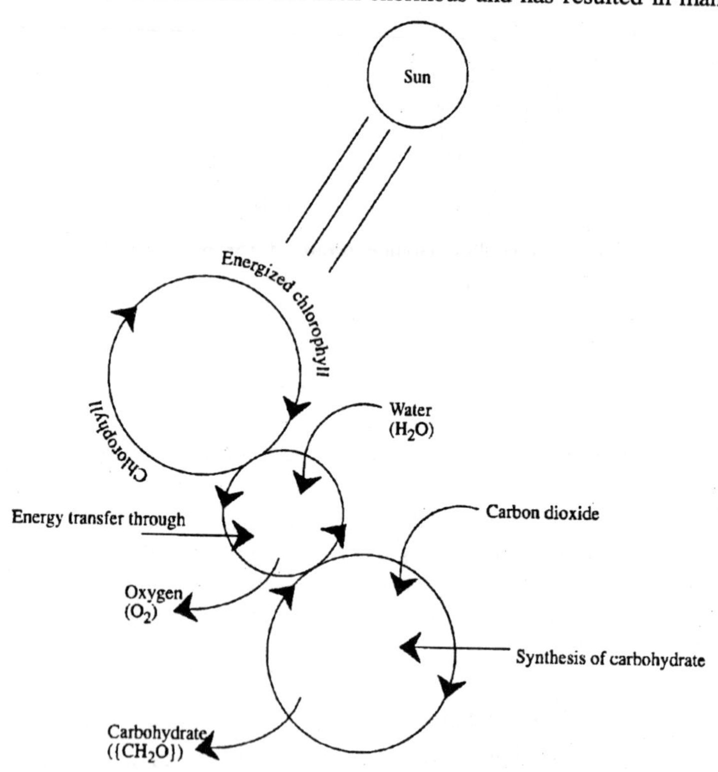

Fig. 1.3. Energy conversion and transfer by photosynthesis.

of the environmental problems now facing humankind. This time period has seen a transition from the almost exclusive use of energy captured by photosynthesis and utilized as biomass (food to provide muscle power, wood for heat) to the use of fossil fuels for about 90 percent, and nuclear energy for about 5 percent of all energy employed commercially. Fossil fuel consumption is divided primarily among petroleum, natural gas, and coal. These sources of energy are limited and their pollution potential is high. The mining of coal and the extraction of petroleum is environmentally disruptive, the combustion of high-sulfur coal releases acidic sulfur dioxide to the atmosphere, and all fossil fuels produce carbon dioxide, a greenhouse gas. Therefore, it will be necessary to move toward the utilization of alternate energy sources, particularly those that are renewable and that do not pose unacceptable risks to global climate through emission of carbon dioxide. Prominent among these is solar energy, and biomass will to a degree come back as an energy source.

HUMAN IMPACT AND POLLUTION

The demands of increasing population coupled with the desire of most people for a higher material standard of living are resulting in worldwide pollution on a massive scale. Environmental pollution can be divided among the categories of water, air, and land pollution. All three of these areas are linked. For example, acid gases emitted to the atmosphere can be converted to strong acids by atmospheric chemical processes, fall to the earth as acid rain, and pollute water with acidity. Improperly discarded hazardous wastes can leach into groundwater that is eventually released as polluted water into streams.

Water Pollution

Throughout history, the quality of drinking water has been a factor in determining human welfare. Waterborne diseases in drinking water have decimated the populations of whole cities. Unwholesome water polluted by natural sources has caused great hardship for people forced to drink it or use it for irrigation. Although waterborne disease have in general been well controlled in industrialized nations, they are prevalent, and even growing worse in poorer countries, especially where population pressures have overtaxed the resources available to provide safe drinking water and to treat wastewater. Currently, toxic chemicals pose the greatest threat to the safety of water supplies in industrialized nations.

Since World War II there has been a tremendous growth in the manufacture and use of synthetic chemicals. Many of the chemicals have contaminated water supplies. Two examples are insecticide and herbicide runoff from agricultural land, and industrial discharge into surface waters. Another serious problem is the threat to groundwater from waste chemical dumps and landfills, storage lagoons, treating ponds, and other facilities. It is clear that water pollution should be a concern of every citizen. Understanding the sources, interactions, and effects of water pollutants is essential for controlling pollutants in an environmentally safe and economically acceptable manner. Above all, an understanding of water pollution and its control depends upon a basic knowledge of aquatic environmental science.

Air Pollution

Inorganic air pollutants consist of many kinds of substances. Many solid and liquid substances may become particulate air contaminants. Another important class of inorganic air pollutants consists of oxides of carbon, sulfur, and nitrogen. Carbon monoxide is a directly toxic material that is fatal at relatively small doses. Carbon dioxide is a natural and essential constituent of the atmosphere, and it is required for plants to use during photosynthesis. However, CO_2 may turn out to be the most deadly air pollutant of all because of its potential as a greenhouse gas that might cause devastating global warming. Oxides of sulfur and nitrogen are acid-forming gases that can cause acid precipitation. Ammonia, hydrogen, chloride, and hydrogen sulfide, are also inorganic air pollutants.

A number of gaseous inorganic pollutants enter the atmosphere as the result of human activities. Those added in the greatest quantities are carbon monoxide (CO), sulfur dioxide (SO_2), nitric oxide (NO) and nitrogen dioxide (NO_2). other inorganic pollutant gases include ammonia, (NH_3), nitrous oxide (N_2O) hydrogen sulfide (H_2S), elemental chlorine (Cl_2), hydrogen chloride (HCl), and hydrogen fluoride (HF). Substantial quantities of some of these gases are added to the atmosphere each year by human activities. Globally, atmospheric emissions of carbon monoxide, sulfur oxides, and nitrogen are of the order of one to several hundred million tons per year.

Organ pollutants are common atmospheric contaminants that may have a strong effect upon atmospheric quality. Such pollutants may come from both natural and artificial sources. In some cases contaminants from both kinds of sources interact to produce a pollution effect. This occurs, for example, when terpene hydrocarbons evolved from citrus and conifer trees interact with nitrogen oxides from automobiles to produce photochemical smog.

The effects of organic pollutants in the atmosphere may be divided into two major categories. The first consists of *direct effects*, such as cancer caused by exposure to vinyl chloride. The second is the formation of *secondary pollutants*, especially photochemical smog. In the case of pollutant hydrocarbons in the atmosphere, the latter is the more important effect. In some localized situations, particularly the workplace, direct effects of organic air pollutants may be equally important.

Pollution of the Geosphere and Hazardous Waste

The most serious kind of pollutant that is likely to contaminate the geosphere, particularly soil, consists of hazardous wastes. A simple definition of a *hazardous waste* is that it si a hazardous substance that has been discarded, abandoned, neglected, released, or designated as a waste material, or one that may interact with other substances to be hazardous. In a simple sense a hazardous waste is a material that has been left where it may cause harm if encountered.

Humans have always been exposed to hazardous substances going back to prehistoric times when they inhaled noxious volcanic gases or succumbed to carbon monoxide from inadequately vented fires in cave dwellings sealed too well against Ice-Age cold. As the production of dyes and other organic chemicals developed from the coal tar industry in Germany during the 1800s, pollution and poisoning from coal tar by-products was observed. By around 1900 the quantity and variety of chemical wastes produced each year was increasing sharply with the addition of wastes such as spend steel and iron pickling liquor, lead battery wastes, chromic wastes, petroleum refinery wastes, radium wastes, and fluoride wastes from aluminum ore refining. As the century progressed into the World War II era, the wastes and hazardous by-products of manufacturing increased markedly from sources such as chlorinated solvents manufacture, pesticides synthesis, polymers manufacture, plastics, paints, and wood preservatives. The Love Canal affair of the 1970s and 1980s brought hazardous wastes to the public attention as a major political issue in the U.S. Starting around 1940, large quantities of at least 80 different waste chemicals were dumped into this old abandoned canal in Niagara Falls, New York. Serious problems developed that required massive remedial action, and by 1993 state and federal governments had spent well over $100 million to clean up the site and relocate residents.

Technology and the Problems it Poses

Modern technology has provided the means for massive alteration of the environment and pollution of the environment. As discussed in the following section, technology, intelligently applied with a strong environmental awareness, also provides the means for dealing with problems of environmental pollution and degradation.

Some of the major ways in which modern technology has contributed to environmental alteration and pollution are the following:

1. Agricultural practices that have resulted in intensive cultivation of land, drainage of wetlands, irrigation of arid lands, and application of herbicides and insecticides.
2. Manufacturing of huge quantities of industrial products that consumes vast amounts of raw materials and produces large quantities of air pollutants, water pollutants, and hazardous waste by-products.
3. Extraction and production of minerals and other raw materials with accompanying environmental disruption and pollution.
4. Energy production and utilization with environmental effects that include disruption of soil by strip mining, pollution of water by release of salt-water from petroleum production, and emission of air pollutants, such as acid-rain-forming sulfur dioxide.
5. Modern transportation practices, particularly reliance on the automobile, that cause scarring of land surfaces from road construction, emission of air pollutants, and greatly increased demands for fossil fuel resources.

Solutions Offered by Science and Technology

Technology based on a firm foundation of environmental science can be very effectively applied to the solution of environmental problems. One important example of this is the redesign of basic manufacturing processes to minimize raw material consumption, energy use, and waste production. With proper design such a process can be made environmentally relatively more acceptable. In some cases raw materials and energy sources can be chosen in ways that minimize environmental impact. If the process involves manufacture of a chemical, it may be possible to completely alter the reactions used so that the process is much more environmentally friendly. Raw materials and water may be recycled to the maximum extent possible. Best available technologies may be employed to minimize air, water, and solid waste emissions. Among the ways in which technology can be applied to minimize environmental impact are the following:

1. Use of state-of-the-art computerized control to achieve maximum energy efficiency, maximum utilization of raw materials, and minimum production of pollutant by-products.

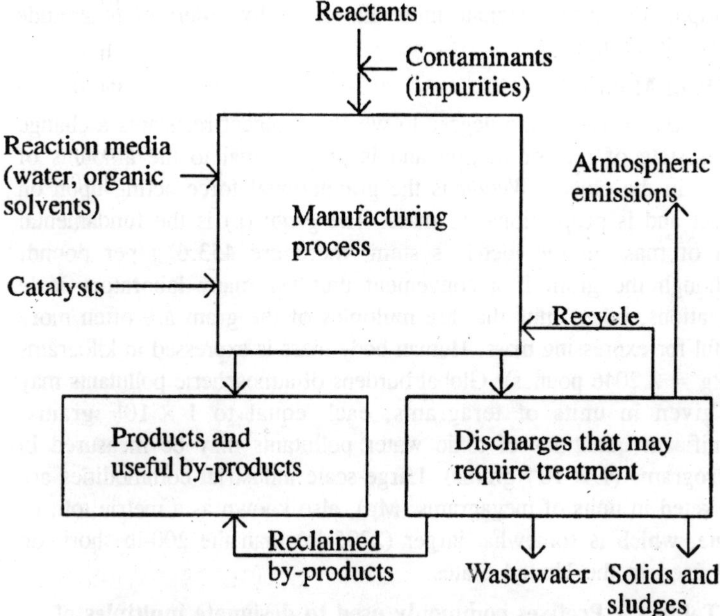

Fig. 1.4. A manufacturing process viewed from the standpoint of minimization of environmental impact.

2. Use of materials that minimize pollution problems, for example, heat-resistant materials that enable use of high temperatures for efficient thermal processes.
3. Application of processes and materials that enable maximum materials recycling and minimum waste product production, for example, advanced membrane processes for wastewater treatment to enable water recycling.
4. Application of advanced biotechnologies, such as in the biological treatment of wastes.
5. Use of best available catalysts for efficient synthesis.
6. Use of lasers for precision machining and processing to minimize waste production.

UNITS

Already in this chapter quantities have been expressed in several units of various kinds. A basic knowledge of units and how to use them is essential for any of the sciences.

The *metric* system has long been the standard system for scientific measurement and is the one most commonly used in this book. It uses multiples of 10 to designate units that differ by orders of magnitude from a basic unit.

Units of Mass

Mass expresses the degree to which an object resistants a change in its state of rest or motion and is proportional to the amounts of mater in the object. *Weight* is the gravitational force acting upon on object and is proportional to mass. The *gram* (g) is the fundamental unit of mass in the metric system; there are 453.6 g per pound. Although the gram is a convenient unit for many laboratory-scale operations, other units that are multiples of the gram are often more useful for expressing mass. Human body mass is expressed in kilograms (1 kg = 2.2046 pounds). Global burdens of atmospheric pollutants may be given in units of teragrams, each equal to 1×10^{12} grams. Significant quantities of toxic water pollutants may be measured in micrograms (1×10^{-6} grams). Large-scale industrial commodities are marketed in units of megagrams (Mg), also known as a metric ton, or tonne, which is somewhat larger (2205 lb) than the 200-lb short ton still used in the United States.

Table 1.1 Prefixes commonly used to designate multiples of unites.

Prefix	*Basic unit is multiplied by*	*Abbreviation*
Mega	1,000,000 (10^6)	M
Kilo	1,000 (10^3)	k
Hecto	100 (10^2)	h
Deka	10 (10)	da
Deci	0.1 (10^{-1})	d
Centi	0.01 (10^{-2})	c
Milli	0.001 (10^{-3})	m
Micro	0.000,001 (10^{-6})	μ
Nano	0.000,000,001 (10^{-9})	n
Pico	0.000,000,000001 (10^{-12})	p

Units of length

Length in the metric system is expressed in units based upon the *meter*, m; a meter is 39.37 inches long, slightly longer than a yard. A kilometer (km) is equal to 1000 m and, like the mile, is used to measure relatively great distances. A centimeter (cm) equal to 0.01 m is often convenient to designate lengths such as dimensions of laboratory instruments. There are 2.540 cm per inch, and the cm is employed to express lengths that would be given in inches in the English system. The micrometer (μm) is about the same length as that of a typical bacterial cell. The μm is also used to express wavelengths of infrared radiation by which Earth re-radiates solar energy back to outer space. The nanometer (nm), equal to 10^{-9} m, is a convenient unit for the wavelength of visible light, which ranges from 400 to 800 nm.

Units of volume

The basic metric unit of *volume* is the liter. This volume is defined in terms of metric units of length. A liter is the volume of a decimeter cubed, that is, 1L = 1 dm^3 (a decimeter is 0.1 meter, about 4 inches), and is equal to 1.057 quarts. A milliliter (mL) is the same volume as a cubic centimeter, cm^3.

Units of temperature

In science, temperatures may be expressed in metric units of *Celsius degrees*, °C. On this scale, water freezes at 0°C and boils at 100°C. The most fundamental temperature is the *Kelvin* or *absolute* scale, for which zero is the lowest attainable temperature. A unit of temperature on this scale is equal to a Celsius degree, but it is called a *Kelvin*, not a degree, and is designated as K, not °K. The value of absolute zero on the Kelvin scale is –273.15°C, so that the Kelvin temperatures is always a number 273.15 (usually rounded to 273) higher than the Celsius temperature. Thus water boils at 373 K and freezes at 273 K.

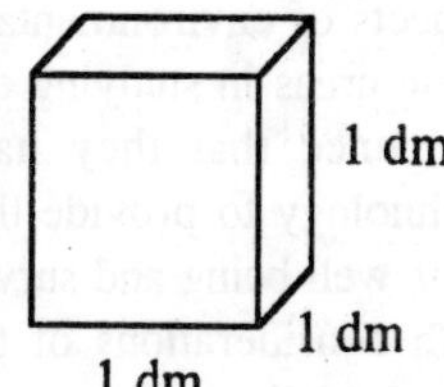

Fig. 1.5. A cube that is 1 decimeter to the side has a volume of 1 liter.

Units of pressure

Pressure is force per unit area and can be expressed in a number of different units, including the *atmosphere* (atm), which is the average pressure exerted by air at sea level, or the pascal (Pa), usually expressed in kilopascal (1 kPa = 1000 Pa, and 101.3 kPa = 1 atm). Based on pressure required to hold up a column of mercury in a mercury barometer, pressure can be given as ***millimeters of mercury*** (mm Hg) where 1 mm of mercury is a unit called the *torr* and 760 torr equal 1 atm.

2

ENVIRONMENT AND THE ANTHROSPHERE

This chapter addresses the technological, engineering, and industrial aspects of environmental science. It is absolutely essential to consider these areas in studying environmental science because of the enormous influence that they have on the environment. Humans will use technology to provide the food, shelter, and goods that they need for their well-being and survival. The challenge is to interweave technology with considerations of the environment and ecology such as the two are mutually advantageous.

Technology, properly applied, is an enormously positive influence for environmental protection. The most obvious such application is in air and water pollution control. Necessary as "end-of-pipe" measures are not the control of air and water pollution, it is much better to use technology in manufacturing processes to prevent the formation of pollutants. Technology is being used increasingly to develop highly efficient processes of energy conversion, renewable energy resource utilization, and conversion of raw materials to finished goods. In the transportation area, properly applied technology in areas such as high speed train transport can enormously increase the speed, energy efficiency, and safety of means for moving people and goods.

WHERE HUMANS LIVE

The dwellings of humans have an enormous influence on their well-being and on the surrounding environment. In relatively affluent societies the quality of living space has improved dramatically during the last century. Homes have become much more spacious per occupant and have become largely immune to the extremes of weather conditions. Such homes are equipped with a huge array of devices, such as indoor plumbing, climate control, communications equipment, and entertainment centers. The comfort factor for occupants has increased enormously.

The construction and use of modern homes and the other buildings in which people spend most of their time place tremendous strains on their environmental support systems and cause a great deal of environmental damage. Typically, as part of the siting and construction of new homes, shopping centers, and other buildings, the landscape is rearranged drastically at the whims of developers. Top soil is removed, hills are cut down, and low places are filled in an attempt to make the surrounding environment fit to a particular architectural scheme. The construction of modern buildings consumes large amounts of resources, such as concrete, steel, plastic, and glass, as well as the energy required to make synthetic building materials. The operation of a modern building requires additional large amounts of energy and of materials, such as water. It has been pointed out that all too often the design and operation of modern homes and other buildings takes place "out of the context" of the surroundings and the people who must work in and occupy the buildings.

Although the above discussion of the environmental aspects of modern buildings can be construed as largely "bad news," the "good news" is that there is a large potential to design, construct, and operate hoes and other buildings in a manner consistent with environmental preservation and improvement. One obvious way in which this can be done is to reevaluate the kinds of materials used in buildings. The manufacture of steel, for example, requires mining iron ore and coking coal, consumption of the energy required to make the steel, and dealing with the emissions and wastes associated with steel production. Similarly, the synthesis and fabrication of plastic used in buildings requires large amounts of petroleum and energy. Substitution of renewable materials, such as wood, and nonfabricated materials, such as quarried stone, can save large amounts of energy and minimize environmental impact. In some parts of the world sun-dried adobe blocks made from soil are practical building materials that require little energy to fabricate.

Recycling of building materials and of whole buildings can save large amounts of materials and minimize environmental damage. At a low level, stone, brick, and concrete can be used as fill material upon which new structures may be constructed. Bricks are often recyclable, and recycled used bricks often make useful and quaint materials for walls and patios. Given careful demolition practices, wood can often be recycled. Buildings can be designed with recycling in mind. This means using architectural design conducive to adding stories and annexes and to rearranging existing space. Utilities may be placed in readily accessible passageways rather than being imbedded in structural components in order to facilitate later changes and additions.

Technological advances can be used to make buildings much more environmentally friendly. Advanced window design using multiple panes and infrared-blocking glass can significantly reduce energy consumption. Modern insulation materials are highly effective. Advanced heating and air conditioning systems operate with a high degree of efficiency. Automated and computerized control of building utilities, particularly those used for cooling and heating, can significantly reduce energy consumption by regulating temperatures and lighting to the desired levels at specific locations and times in the building.

Advances in making buildings airtight and extremely well insulated can lead to problems with indoor air quality. Carpets, paints, panelling, and other manufactured components of buildings give off organic vapors, such as formaldehyde, solvents, and monomers used to make plastics and fabrics. In a poorly insulated building that is not very airtight such indoor air pollutants cause few if any problems for the building occupants. However, extremely airtight buildings can accumulate harmful levels of indoor air pollutants. Therefore, building design and operation to minimize accumulation of toxic indoor air pollutants is receiving a much higher priority.

How Humans Move

Few aspects of modern industrialized society have had as much influence on the environment as developments in transportation. These effects have been both direct and indirect. The direct effects are those resulting from the construction and use of transportation systems. The most obvious example of this is the tremendous effects that the widespread use of automobiles, trucks, and buses have had upon the environment. Entire landscapes have been entirely rearranged to construct highways, interchanges, and parking lots. Emissions from the internal combustion engines used in automobiles are the major source of air pollution in many urban areas.

The indirect environmental effects of widespread use of automobiles are enormous. The automobile has made possible the "urban sprawl" that is characteristic of residential and commercial patterns of development in the U.S., and in many other in industrialized countries as well. The paving of vast areas of watershed and alteration of runoff patterns have contributed to flooding and water pollution. Discarded, worn-out automobiles have caused significant waste disposal problems. Vast enterprises of manufacturing, mining, and petroleum production and refining required to support the "automobile habit" have been very damaging to the environment.

On the positive side, however, applications of advanced engineering and technology to transportation can be of tremendous benefit to the environment. Modern rail and subway transportation systems, concentrated in urban areas and carefully connected to airports for longer distance travel, can enable the movement of people rapidly, conveniently, and safely, with minimum environmental damage. Although pitifully few in number in respect to the need for them, examples of such systems are emerging in progressive cities, showing the way to environmentally friendly transportation systems of the future.

Telecommuter Society

A new development that is just now beginning to reshape the way humans move, where they live, and how they live, is the growth of a *telecommuter society*, composed of workers who do their work at home and "commute" through their computers, modems, FAX machines, and the internet connected by way of high speed telephone communication lines. These new technologies, along with several other developments in modern society, have made such a work pattern possible and desirable. An increasing fraction of the work force deals with information in their jobs. In principle, information can be handled just as well from a home office as it can from a centralized location, often an hour or more commuting distance from the worker's dwelling.

Actually, home and its immediate surroundings were where most work was done prior to the industrial revolution, whose assembly lines and large centralized factories demanded that workers come to a particular location for their work shifts. This work pattern and the prosperity that it brought with it resulted in the establishment of a huge sub urban population that had left the cities and farms. The idyllic dream of suburbia has all too often given way to urban sprawl, traffic congestion, and long, tedious commutes. The associated environmental problems have been enormous.

Within the next approximately 10 years, it is estimated that almost 20% of the U.S. work force, a total of around 30 million people, may be working out of their homes. This tendency has been accelerated in recent years by the downsizing of many U.S. industrializes and 'out sourcing" of work to private parties, sometimes from the companies to the workers that have been laid off. There are disadvantages, of course. People working at home may become isolated, careless in their habits, and "desocialized" from the workplace environment and the "office policy" that can play a role in advancement and training. Indeed, it is possible to visualize another type of home workplace activity in which "telecommuter counselors" advise home workers about how to manage such problems.

The changes that will result from widespread use of telecommuting are many and profound. With properly sited housing, workers can live in a rural environment with minimal disturbance to the surroundings. The flow of information has essentially no environmental costs compared to the movement of people on daily long commutes. The potential benefits to family life are obvious and may play a strong role in reshaping society positively. Whole communities may be transformed by telecommuters. Rather than being populated by people who are gone most of the day to a job far away, telecommuter communities well be occupied by people who are gone most of the day to a job far away, telecommuter communities will be occupied by people who are available on flexible schedules during the day and who are not too exhausted by long, tiring commutes to engage in civic activities at night. Telecommuters are likely to be intelligent, ambitious, and self-motivated, with all that these characteristics imply for civic activities. Cultural activities should thrive in environmentally friendly telecommuter communities combining the best of rural, suburban, and urban life.

How Humans Communicate

It has become on overworked cliche that we live in an information age. Nevertheless, the means to acquire, store, and communicate information are expanding at an incredible pace. This phenomenon is having tremendous effect upon society and has the potential to have numerous effects upon the environment.

The major areas to consider in respect to information are its acquisition, recording, computing, storing, displaying, and communicating. Consider, for example, the detection of a pollutant in a major river. Data pertaining to the nature and concentration of the pollutant may be obtained with a combination gas chromatograph and mass spectrometer. Computation by digital computer is employed to determine the identity and concentration of the pollutant. The data can be stored on a magnetic disk, displayed on a video screen, and communicated instantaneously all over the world by satellite and fiber optic cable.

All the aspects of information and communication listed above have been tremendously augmented by recent technological advances. Perhaps the greatest such advances has been that of silicon integrated circuits. Optical memory consisting of information recorded and ready by microscopic beams of laser light has enabled the storage of astounding quantities of information on a single compact disk. The use of optical fibers to transmit information digitally by light has resulted in a comparable advance in the communication of information.

The central characteristic of communication in the modern age is the combination of telecommunications with computers called *telematics*. Automatic teller machines use telematics to make cash available to users at locations far from the customer's bank. Information used for banking, for business transactions, and in the media depends upon telematics.

There exists a tremendous potential for good in the applications of the "information revolution" to environmental improvement. An important advantage is the ability to acquire, analyze, and communicate information about the environment. For example, such a capability enables detection of perturbations in environmental systems, analysis of the data to determine the nature and severity of the pollution problems causing such perturbations, and rapid communication of the findings to all interested parties.

WHAT HUMANS CONSUME

Food

The most basic human need is the need for food. Without adequate supplies of food, the most pristine and beautiful environment becomes a hostile place for human life. The industry that provides food is *agriculture*, an enterprise concerned primarily with growing crops and livestock.

The environmental impact of agricultural is enormous. One of the most rapid and profound changes in the environment that has ever taken place was the conversion of vast areas of the North American continent from forests and grasslands to cropland. Throughout most of the continental United States, this conversion took place predominantly during the 1800s. The effects of it were enormous. Huge acreages of forest lands that had been stable since the last Ice Age were suddenly derived of stabilizing tree cover and subjected to water erosion. Prairie lands put to the plow were destabilized and subjected to extremes of heat, drought, and wind that resulted in the blowing away of topsoil, culminating in the Dust Bowl of the 1930s.

In recent decades, valuable farmland has faced a new threat posed by the urbanization of rural areas. Prime agricultural land has been turned into subdivisions and paved over to create parking lots and streets. Increasing urban sprawl has led to the need for more highways. In a vicious continuing circle, the availability of new highway system has enabled even more development. The ultimate result of this pattern of development has been the removal of once productive farmland form agricultural use.

On a positive note, agriculture has been a sector in which environmental improvement has seen some notable advances during the last 50 to 75 years. This has occurred largely under the umbrella of soil conservation. The need for soil conservation became particularly obvious during the Dust Bowl years of the 1930s, when it appeared that much of the agricultural production capacity of the U.S. would be swept away from drought-stricken soil by erosive winds. In those times and areas in which wind erosion was not a problem, water erosion took its toll. Ambitious programs of soil conservation have largely alleviated these problems. Wind erosion has been minimized by practices such as low-tillage agriculture, strip cropping in which crops are grown in strips alternating with strips of summer-fallowed crop stubble, and reconversion of marginal cultivated land to pasture. The application of low-tillage agriculture and the installation of terraces and grass waterways have greatly reduced water erosion.

Food production and consumption are closely linked with industrialization and the growth of technology. It is an interesting observation that those countries that develop high population densities prior to major industrial development experience two major changes that strongly impact food production and consumption.

1. Cropland is lost as a result of industrialization; if the industrialization is rapid, increases in grain crop productivity cannot compensate fast enough for the loss of cropland to prevent a significant fall in production.
2. As industrialization raises incomes, the consumption of livestock products increases, such that demand for grain to produce more meat, milk, and eggs rises significantly.

To date, the only three countries that have experienced rapid industrialization after achieving a high population density are Japan, Taiwan, and South Korea. In each case, starting as countries that were largely self sufficient in grain supplies, these nations lost 20-30 percent of their grain production and became heavy grain importers over an approximately three-decade time period. The effects of these changes on global rain supplies and prices was relatively small because of the limited population of these countries—the largest, Japan, had a population of only about 100 million.

Since approximately 1990, however, China has been experiencing economic growth at a rate of about 10% per year. With a population of 1.2 billion people, China's economic activity has an enormous effect on global markets. It may be anticipated that this economic growth, coupled with a projected population increase of more than 400 million people during the next 30 years, will result in a demand for grain and other food supplies that will cause disruptive food shortages and dramatic price increases. As an indication of what might happen, China experienced a 24 percent rate of inflation during 1994, largely due to higher prices for scarce food commodities.

In addition to the destruction of farmland to build factories, roads, housing and other parts of the infrastructure associated with industrialization, there are other factors that tend to decrease grain production as economic activity increases. One of the major ones of these is air pollution, which decreases crop production. Water pollution can seriously curtail fish harvests. Intensive agriculture uses large quantities of water for irrigation. If groundwater is used for irrigation, aquifers may become rapidly depleted.

The discussion above points out several factors that are involved in supplying food to a growing world population. There are numerous complex interactions among the industrial, societal, and agricultural sectors, Changes in one inevitably result in changes in other sectors.

Infrastructure

The *infrastructure* refers to the utilities, facilities, and systems used in common by members of a society and upon which the society is dependent for its normal function. The infrastructure includes both physical components—roads, bridges, and pipelines—and the instructions—laws, regulations, and operational procedures—under which the physical infrastructure operates. Parts of the infrastructure may be publicly owned, such as the U.S. Interstate Highway system and some European railroads, or privately owned, as is the case with virtually all railroads in the U.S. Some of the major components of the infrastructure of a modern society are the following:

1. Transportation systems, including railroad, highways, and air transport systems.
2. Energy generating and distribution systems.
3. Buildings
4. Telecommunications systems
5. Water supply and distribution systems
6. Waste treatment and disposal systems, including those for municipal wastewater, municipal solid refuse, and industrial wastes.

In general, the infrastructure refers to the facilities that large segments of a population must use in common in order for a society to function. In a sense, the infrastructure is analogous to the operating system of a computer. A computer operating system determines how individual applications operate and the manner in which they distribute and store the documents, spreadsheets, and illustrations created by the applications. Similarly, the infrastructure is used to move raw materials and power to factories and to distribute and store their output. An outdated, cumbersome computer operating system with a tendency to crash is detrimental to the efficient operation of a computer. In a similar fashion an outdated, cumbersome, broken down infrastructure causes society to operate in a very inefficient manner and is subject to catastrophic failure.

For a society to be successful it is of the utmost importance to maintain a modern, viable infrastructure. Such an infrastructure is consistent with environmental protection. Properly designed utilities and other infrastructure elements, such as water supply systems and wastewater treatment systems, minimize pollution and environmental damage. Components of the infrastructure are subject to deterioration. To a large extent this is due to natural aging processes. Fortunately, many of these processes can be slowed or even reversed. Corrosion of steel structures, such ad bridges, is a big problem for infrastructure; however, use of corrosion-resistant materials and maintenance with corrosion-resistant coatings can virtually stop this deterioration process. The infrastructure is subject to human insult, such as vandalism, misuse, and neglect. Often the problem begins with the design and basic concept of a particular component of the infrastructure. For example, many river dikes destroyed by flooding should never have been built because they attempt to thwart to an impossible extent the natural tendency of rivers to flood periodically.

Technology plays a major role in building and maintaining a successful infrastructure. Many of the most notable technological advances applied to the infrastructure were made from 150 to 100 years ago. By 1900 railroads, electric utilities, telephones, and steel building skeletons had been developed. The net effect of most of these technological innovations was to enable humankind to "conquer," or at least temporarily subdue nature. The telephone and telegraph helped to overcome isolation, high speed rail transport and later air transport conquered distance, and dams were used to control rivers and water flow.

The development of new and improved materials is having a significant influence on the infrastructure. From about 1970 to 1985 the strength of steel commonly used in construction nearly doubled. During the later 1900s significant advances were made in the properties of structural concrete. Superplasticizers enables mixing cement with less water, resulting in a much less porous, stronger concrete product. Polymeric and metallic fibers used in concrete made it much stronger. For dams and other applications in which a material stronger than earth but not as conventional concrete is required, roller-compacted concrete consisting of a mixture of cement with silt or clay has been found to be useful. The silt or clay used is obtained on site with the result that both construction costs and times are lowered.

The major challenge in designing and operating the infrastructure in the future will be to use it to work with the environment and to enhance environmental quality to the benefit of humankind. Obvious examples of environmentally friendly infrastructures are state-of-the-art sewage treatment systems, high-speed rail systems that can replace inefficient highway transport, and stack gas emission control systems in power plants. More subtle approaches with a tremendous potential for making the infrastructure more environmentally friendly include employment of workers at a computer terminal in their homes to that they do not need to commute, instantaneous electronic mail that avoids the necessity for moving letters physically, and solar electric powered installations to operate remote signals and relay stations, which avoids having to run electric power lines to them.

Whereas advances in technology and the invention of new machines and devices enabled rapid advances in the development of the infrastructure during the 1800s and early 1900s, it may be anticipated that advances in electronics and computers will have a comparable effect in the future. One of the areas in which the influence of modern electronics and computers is most visible is in telecommunications. Dial telephones and mechanical relays were perfectly satisfactory in their time, but have been made totally obsolete by innovations in electronics, computer control, and fiber optics. Air transport controlled by a truly modern, state-of-the-art computerized control system (which, unfortunately, is not yet fully installed in the U.S.) could enable present airports to handle many more airplanes safely and efficiently, thus reducing the need for airport construction. Sensors for monitoring strain, temperature, movement, and other parameters can be imbedded in the structural members of bridges and other structures. Information from thee sensors can be e processed by computer to warn of failure and to aid in proper maintenance. Many similar examples could be cited.

Although the payoff is relatively long term, intelligent investment in infrastructure pays very high rewards. In addition to the traditional rewards in economics and convenience, properly designed additions and modifications to the infrastructure can pay large returns in environmental improvement as well.

Technology and Engineering

Technology refers to the ways in which humans do and make things with materials and energy. In the modern era, technology is to a large extent the product of engineering based on scientific principles. Science deals with the discovery, explanation, and development of theories pertaining to interrelated natural phenomena of energy, matter, time, and space. Based on the fundamental knowledge of science, engineering provides the plans and means to achieve specific practical objectives, Technology uses these plans to carry out the desired objectives.

Technology has a long history, and indeed, goes back into pre-history to times when humans used primitive tools made from stone, wood, and bone. As humans settled in cities, human and material resources became concentrated and focused, such that technology began to develop at an alliterating pace. Technology got a tremendous boost from the discovery that metals could be worked and shaped, giving rise to the pursuit of *metallurgy*. Metallurgy probably began with processing of native elemental around 4000 B.C., followed by the widespread use of bronze, an alloy of tin and copper. The Bronze Age lasted until around 1200 B.C., when iron replaced bronze for tools and weapons. Other early technological innovations predating the rise of Greek and Roman civilizations included domestication of the horse, discovery of the wheel, architecture to enable construction of substantial buildings, control of water for canals and irrigation, and writing for communication.

A major advance during the Greek and Roman eras was the development of *machines*, including the windlass, pulley, inclined plane, screw, catapult for throwing missiles in warfare, and water screw for moving water. Later, the water wheel was developed for power, which was transmitted by wooden gears. Many technological innovations came from China, sometimes several hundred or even a thousand years before they appeared in Europe. As examples, China and rotating fans for ventilation by around 200 A.D., bringing with wood blocks by around 740, and gunpowder about a century later. The 1800s saw an explosion in technology. Among the major advances during this century were widespread use of steam power, steam-powered railroad, the telegraph, telephone, electricity as a power source, textiles, use of iron and steel in building and bridge construction, cement, photography, and invention of the internal combustion engine, which revolutionized transportation in the following century. It may be argued that in a relative sense advances in technology during the 1800s were at least as great as those that have occurred since 1900s, and certainly laid the groundwork for the vast advances that took place during the last 100 years.

Since about 1900, advancing technology has been characterized by vastly increased uses of energy; greatly increased speed in manufacturing processes, information transfer, computation, transportation, and communication; automated control; a vast new variety of chemicals; new and improved materials for new applications; and, more recently, the widespread application of computers to manufacturing, communication, and transportation. Arguably, the greatest impact during the last 100 years has been the application of electronics to technology. Electronics as it is now understood began with the discovery of the radio by Guglielmo Marconi in 1896. Electronics got an enormous boost in the early 1900s with the development of vacuum tubes that had the ability to control and amplify electronic signals and to make radio transmission possible over vast distances. A special vacuum tube, the cathode ray tube, made television and radar possible. Solid-state devices that are remarkably small and fast have supplanted vacuum tubes in virtually all applications and have made modern computers possible.

The development of electronics probably illustrates better than anything else the revolution in technology resulting from applications of basic and applied science. In modern times science and technology are inseparable and synergistic. The principles of science are applied to make technological advances possible. For example, radio communication and electronics was made possible by understanding of electromagnetic waves.

In transportation, the development of passenger-carrying airplanes has affected an astounding change in the ways in which people get around. In addition, large amounts of high-priority freight are now moved by air.

The technological advances of the present century are largely attributable to improved materials. For example, since before World War II, airliners have been made of special strong alloys of aluminum; these are being supplanted by even more advanced composites. Synthetic materials with a significant impact on modern technology include plastics, fiber reinforced materials, composites, and ceramics.

Until very recently, technological advances were made largely without heed to environmental impacts. Now, however, the greatest technological challenge is to reconcile technology with environmental consequences. The survival of humankind and of the planet that supports it now requires that the established two-way interaction between science and technology become a three-way relationship including environmental protection. That is, of course, a major theme of this book.

Engineering

Engineering uses fundamental knowledge acquired through science to provide the plans and means to achieve specific objectives in areas such as manufacturing, communication, and transportation. At one time engineering could be divided conveniently between military and civil engineering. With increasing sophistication, civil engineering evolved into even more specialized areas, such as mechanical engineering, chemical engineering, electrical engineering, and environmental engineering. Other engineering specialities include aerospace engineering, agricultural engineering, biomedical engineering, CAD/CAM (computer-aided design and computer-aided manufacturing engineering), ceramic engineering, industrial engineering, materials engineering, metallurgical engineering, mining engineering, plastics engineering, and petroleum engineering.

Mechanical engineering is the branch of engineering that deals with machines and the manner in which they handle forces, motion, and power. This discipline arose as a separate area with the development of the enormous capabilities of the steam engine in the early 1800s. The major objective of mechanical engineering is to develop and improve machines that produce goods and services. The scientific principles upon which engineering is based include consideration of the laws that govern forces and motion (dynamics); the thermodynamic laws that govern energy, power, and heat; transfer of materials and fluids; and other factors, including vibration control, materials properties, and wear minimization through proper lubrication. In addition to mechanical components, other machine components with which mechanical engineering must deal are electric, electronic, hydraulic, and fluidic components.

There are several objectives of machine design. Machines must produce whatever is needed in high quality to minimize expensive rejects. Speed of production is of utmost importance. Finally, costs must be minimized to enable adequate return on capital. An important factor involved in this endeavor is that of materials that can maintain exacting tolerances with minimum wear under sometimes severe conditions. Control of machines is particularly important. This is especially true with greatly increased use of automation and robotics. The availability of inexpensive, fast, sophisticated computers is revolutionizing control of machines and the processes that they carry out.

In the past, many of the machines and processes developed through mechanical engineering have contributed to environmental degradation and pollution. The availability of gargantuan earth-moving equipment has enabled strip-mining, destruction of wildlife habitat, and damming of natural streams. Efficient machine-equipped factories have often produced massive amounts of pollution and have provided noisy and dangerous conditions for workers. As with other branches of engineering, a major emphasis has now been placed on mechanical engineering designed to minimize environmental impact and to improve environmental quality. Examples of this include machinery designed to minimize noise, much improved energy efficiency in machines, and the uses of earth-moving equipment for environmentally beneficial purposes, such as restoration of strip-mined lands and construction of wetlands.

Electrical engineering grew from the rapidly developing electrical power industry starting in the late 1800s. The theory of electrical engineering is largely based on the mathematical formulation of the laws of electricity by the Scotsman James Clerk, Maxwell in 1864, though the profession began in a primitive manner with the invention of the telegraph in 1837. Electrical engineering is concerned with the generation, transmission, and utilization of electrical energy. Of particular importance in this area has been the early utilization of alternating current, which is more complex to use, but ore efficient than direct current.

Proper application of electrical engineering can be very helpful in the environmental area. Efficient generation, distribution, and utilization of electrical energy constitute one of the most promising avenues of endeavor leading to environmental improvement. The modern practice of burying electrical transmission lines in urban areas has minimized visual pollution from unsightly overhead power lines.

Electronics engineering deals with phenomena based on the behavior of electrons in vacuum tubes and other devices. As such, it is very much concerned with electromagnetic radiation ("radio waves" and microwaves) and transmission of radio frequency signals through the air and space. Much of the early work in electronics engineering was based on fundamental studies of the behavior of electrons in vacuum tubes dating from the first decades of the 1900s. Though slow, prone to failure (because of the tendency of their hot filaments to burn out), and power consumptive by modern standards, these devices enabled development of practical radios, television, radar, and chemical instrumentation.

The fact that modern times are largely an "electronic era" is due to the invention of the solid-state electronic device known as the *transistor* in 1948. This enables a truly remarkable development that has changed society enormously in the latter part of this century—the *silicon integrated circuit*. Integrated circuits are made by depositing and etching microscopic electronic circuits containing many transistors, capacitors, and resistors on single silicon chips as small as a pinhead and rarely larger than a dime. Although the cost of integrated circuit chips has remained about constant since 1960, the number of transistors possible on each chip has doubled regularly and may eventually reach a billion or so. This phenomenon has led to the vast array of consumer electronic devices, computers, and control systems that are a fact of modern life—and this is probably only the beginning.

The principles of electronics applied through electronic engineering for environmental improvement are enormous. Automated factories can turn out goods with lowest possible consumption of energy and materials, while minimizing air and water pollutants and production of hazardous wastes. During the last two decades electronic control of electrical power production, distribution, and utilization has enable greater production of light and usable energy without the construction of massive numbers of new power plants. Sophisticated electronic control and the development of electronically-based photovoltaic cells are enabling practical utilizing of solar energy. Nuclear power generation, which can certainly play a major role in generating electrical energy without production of greenhouse gases, can be made virtually fail-safe by electronic systems that do not doze, daydream, or have bad habits that are always potential problems with systems that rely primarily on the performance of fallible humans.

Chemical engineering uses the principles of chemical science, physics, and mathematics to design and operate processes that generate produces and materials through controlled chemical reactions. Historically, a key concept of chemical engineering has bene that of *unit processes*, such as mixing, distillation, filtration, and heat exchange that are combined in a variety of ways to carry out chemical processes. These in turn are based on laws of thermodynamics, chemical kinetic, mass and heat transfer, and fluid flow.

Thermodynamics, the science of heat and energy phenomena, transfers, and conversions, is of crucial importance in chemical engineering. Thermodynamics enables computation of heat required and produced in chemical production, deals with the feasibility of partition of material between phases, and provides information regarding the effects of temperature on chemical equilibrium (degree to which a reaction goes to completion). Whereas thermodynamics deals with equilibrium situations, *chemical kinetics* deals with speeds of chemical reactions. It explains whether or not a reaction proceeds at a sufficient rate to be practical or so rapidly as to be hazardous, and how long reactants must remain together for a desired reaction to occur. A key aspect of chemical kinetics is *catalysis*, in which catalysts, which enable chemical reactions to occur, are either mixed with reactant mixtures (homogeneous catalysis) or held on solid surfaces (heterogeneous catalysis).

Mass transfer involves separation of phases of materials and transfer of materials between phases. The most straightforward examples of mass transfer is distillation in which one component of a mixture is vaporized and condensed. Sorption of pollutant organic matter from aqueous solution onto activated carbon, precipitation and filtration of solids, and extraction of materials from aqueous solution into an organic solvent are all unit operations involving mass transfer. Much of chemical engineering deals with *heat transfer*, which in chemical plants is accomplished by conduction, convection, radiation, and as latent heat (condensation/evaporation, particularly of water).

Unlike other manufacturing and assembly operations that deal with the manipulation of objects and solids, virtually all of the material transferred in a chemical plant is in the form of fluids, so *fluid flow* becomes a major consideration. Consideration must be given to whether or not a fluid can be induced to flow at a sufficiently rapid rate to be practical, as well as what may be done to make a fluid flow adequately. Another major consideration is the loss of energy as fluids flow through pipes, over solid catalysts, and through chemical reactors.

Control of the processes described above is of particular importance in chemical engineering; a poorly regulated process can very rapidly get out of control and ruin the product or cause a fire or explosion. The transfers of fluids and energy that occur in chemical plants are particularly amenable to automated and computerized control.

Acquisition of Raw Materials

Manufacturing involves the processing of a wide variety of materials. Some materials are used with relatively little processing, whereas others are the result of highly sophisticated manufacturing processes. The acquisition and processing of materials has a number of environmental implications. There are numerous potential environmental effects of mining. These include removal and distribution of overburden in strip mining, disturbance of watersheds and aquifers, and release of pollutants to waterways. On the other hand, the development of sophisticated materials through the application of materials science has the potential to be very beneficial to the environment. As an example, underground fuel storage tanks made of polymer-reinforced fiberglass are corrosion proof and will not leak unless subjected to drastic physical insult. Their use virtually eliminates leakage, which was a significant problem with older steel tanks and makes it unnecessary to use anticorrosive coatings, which themselves pose a potential for soil and water pollution.

Raw Materials

Raw materials consist of the minerals, fuel, wood, fire, and other substances required for manufacturing processes. Such materials can be obtained from either *extractive* (nonrenewable) or from *renewable* sources.

Extractive sources of raw materials are those in which substances are dug or pumped from the Earth's crust. Such materials are irreplaceable, so that their wise use and conservation are crucial to the well-being of future generations. They include both inorganic materials and organic substances. Common examples of the former are iron ore, sulfur, and phosphate, whereas examples of organic materials extracted from Earth are coal, natural gas, and crude oil.

The prime example of a renewable material is wood. From an environmental viewpoint wood is an ideal resource. Forests conserve soil by preventing soil erosion and provide recreational areas and watersheds. The removal of carbon dioxide from air by photosynthesis carried out by trees is a significant mechanism for the removal of greenhouse-gas carbon dioxide from the atmosphere. Natural rubber and cotton are two other examples of renewable resources. The use of these substances saves irreplaceable petroleum that otherwise would be consumed in manufacturing synthetic rubber and synthetic fabrics.

Unfortunately, many resources are simply not renewable. It is impossible to grow aluminum, nickel, phosphorus, sulfur, or any of the other essential elemental resources. For these irreplaceable resources, use minimization, substitution by more abundant alternate resources, conservation, and recycling are of utmost importance.

Manufactured Materials

To an increasing degree manufactured materials with special properties are being used for structures, machines and other applications. The use of such materials has both positive and negative aspects from the environmental viewpoint. For example, most such manufactured materials require significant amounts of energy and may sue irreplaceable petroleum as a raw material. On the other hand, some such materials are much more enduring and suitable than those which they replace.

The study of the synthesis, composition, properties and applications of manufactured materials is part of the discipline of *materials science*. Materials science involves a wide variety of materials. Prominent among these are polymers, which are made of large macromolecules synthesized from smaller monomer molecules. There are natural polymeric materials, such as cotton and wood, and synthetic polymers, including polyethylene, nylon, and dacron. Ceramics include a variety of inorganic materials that usually contain silica on and oxygen and are generally formed by high temperature processes. A fast-growing area of materials science deals with composites consisting of two or more materials, often quite dissimilar in their properties, bound together to provide a material with properties superior to those of its separate constitutes. Generally composites consist of reinforcing materials, such as glass fibers or metal wires, bound in some sort of moldable matrix, such as plastic or ceramic.

MANUFACTURING

An *industry* is an enterprise that makes a kind of goods or provides a particular service needed by humans for their existence and well-being. Various industries have developed because specialization of human activities makes for the greatest efficiency in providing goods and services. The kinds of industries that a country or region has demand upon the availability of needed attributes, such as raw materials, human resources, or availability of transport.

Classification of Industries

Industries fall into various classes. In an early stage of development, *basic need industries* providing essentials of food, clothing, fuel, and shelter are emphasized, but become less important relative to more discretionary industries as wealth increases. The most common example of this is the high percentage of people engaged in agriculture at early stages of development, which dwindles to a very low figure (currently only about 3% in the U.S.) as the industrial and economic base becomes more developed. Somewhat arbitrarily, industries may be divided among the following categories:

1. *Food production*. Agriculture and fishing.
2. *Extractive mineral industries* consisting of those involved with the mining of minerals, such as those used as sources of metals (energy sources are addressed in separate categories here).
3. *Renewable resource industries*. Forestry, production of nonfood crops, such as cotton.
4. *Renewable energy industry*, a small, but of necessity, growing industry dealing with the utilization of renewable energy resources, such as solar energy, wind power, and biomass energy.
5. *Extractive energy industry* consisting of coal mining, uranium ore mining, petroleum, and natural gas.
6. *Manufacturing*. Conversion of raw materials or articles to higher-value goods.
7. *Construction*. Building and erection of dwellings, buildings, railroads, highways, and other components of the infrastructure.
8. *Utilities*. Electricity distribution systems, natural gas.
9. *Communications*. Telecommunications, media communications.
10. *Transportation*. Rail, highway, air barge, ship.

11. *Wholesale and retail trade*, which provides the interface between the production of goods and their sale for consumption and use.
12. *Finance*. Banks and other entities that provide the financial resources and transactions required for industries and trade.
13. *Services*. Law, medicine, motels, recreation, and many others.
14. *Government*. National, regional (state, provincial), city, and local entities that provide needed services and regulation.

Industrial growth in developing societies can be described by the sequence of (1) traditional society devoted largely to the most fundamental economic needs of food and shelter; (2) preconditions for take-off; (3) take-off, a growth period characterized by heavy investment in industrial development; (4) drive to maturity in which the industrial base becomes mature and well diversified, and (5) age of high mass consumption, which emphasizes high consumption of consumer goods and a large service industry. It may be hoped that a sixth stage will become dominant in which societies achieve equilibrium with the environment and the resources that must sustain it. Such a stage would be characterized by limited consumption of disposable consumer goods, highly efficient utilization of energy, land use consistent with environmental harmony, zero population growth, and a high quality of life.

Manufacturing

Once a device or product is designed and developed, it must be made—synthesized or manufactured. This may consist of the synthesis of a chemical from raw materials, causing of metal or plastic parts, assembly of parts into a device or product, or any of the other things that go into product that is needed in the marketplace.

Manufacturing activities have a tremendous influence on the environment. Energy, petroleum to make petrochemicals, and ores to make metals must be dug from, pumped from, or grown on the ground to provide essential raw materials. The potential for environmental pollution from mining, petroleum production, and intensive cultivation of soil is enormous. Huge land-disrupting factories and roads must be built to transport raw materials and manufactured products. The manufacture of goods carries with it the potential to cause significant air and water pollution and production of hazardous wastes. The earlier in the design and development process that environmental considerations are taken into account, the more "environmentally friendly" a manufacturing process will be.

Automation, Robotics, and Computers in Manufacturing

Three relatively new developments that have revolutionized manufacturing and that continue to do so are automation, robotics, and computers.

Automation

Automation uses mechanical and electrical devices integrated into systems to replace or extend human physical and mental activities. Primitive forms were known in ancient times; and early example consists of float devices used to control water levels in Roman plumbing systems. A key component of an automated system is the *control system*, which regulates the response of components of a system as a function of conditions, particularly those of time or location.

The simplest level of automation is *mechanization*, in which a machine is designed to increase the strength, speed, or precision of human activities. A back hoe for dirt excavation is an example of mechanization. *Open-loop*, *multifunctional* devices perform tasks according to preset instructions, but without any feedback regarding whether or how the task was done. *Closed-loop*, *multifunctional* devices use process feedback information to adjust the process on a continuous basis. The highest level of automation is *artificial intelligence* in which information is combined with simulated reasoning to arrive at a solution to a new problem or perturbation that may arise in the process.

The greatest application of automation is in manufacturing and assembly. One of two major approaches employed is *fixed automation* used for large numbers of repetitive actions over a long period of time in which the mechanical devices and electrical circuitry of the device determine what it does. Such a device has to be rebuilt to change its function in any major respect. *Programmable automation* is a ,ore flexible approach in which instructions to the machine can be varied so that it can perform different functions.

Not all of the effects of automation on society and on the environment are necessarily good. One obvious problem is increased unemployment and attendant social unrest resulting from displaced workers. Another is the ability that automation provides to enormously increase the output of consumer goods at more affordable prices. This capability greatly increases demands for raw materials and energy, putting additional strain on the environment. To attempt to address such concerns by cutting back on automation is unrealistic, so societies must learn to live with it and to use it in beneficial ways. There are many beneficial applications of technology. Automated processes can result in much more efficient utilization of energy and materials for production, transportation, and other human needs. A prime example is the greatly increased gasoline mileage achieved during the last approximately 20 years by the application of computerized, automated control of automobile engines. Automation in manufacturing and chemical synthesis is used to produce maximum product from minimum raw material. Production of air and water pollutants and of hazardous wastes can be minimized by the application of automated processes. By replacing workers in dangerous locations, automation can contribute significantly to workers health and well-being.

Robotics

Robotics refers to the use of machines to simulate human movements and activities. *Robots* are machines that perform such functions using computer-driven mechanical components to grip, move, reorient, and manipulate objects. Modern robots are characterized by intricately related mechanical, electronic, and computational system. A robot can perform a variety of functions according to pre-programmed instructions that can be changed according to human direction or in response to changed circumstances.

There are a variety of mechanical mechanisms associated with robots.. These are servo mechanisms in which low-energy signals from electronic control devices are used to direct the actions of a relatively large and powerful mechanical system. Robot arms may bend relative to each other through the actions of flexible joints. Specialized end effectors are attached to the ends of robot arms to accomplish specific functions. The most common such device is a gripper used like a hand to grasp objects.

Sensory devices and systems are crucial in robotics to sense position, direction, speed, and other factors required to control the functions of the robot. Sensors maybe used to respond to sound, light, and temperature. One of the more sophisticated types of sensors involves a form of vision. Images captured by video camera can be processed by computer to provide information required by the robot.

Robots interact strongly with their environment. In addition to sensing their surroundings, robots must be able to respond to it in desired ways. In so doing, robots rely on sophisticated computer control. Rapid developments in computer hardware, power, and software continue to increase the ability of robots to interact with their environment. Commonly, instructions to robots are provided by computer programs programmed by humans. It is now possible in many cases to lead a robot through its desired motions and have it "learn" the sequence by computer.

Robots are now used for numerous applications. The main ones of these are for moving materials and objects, performing operations in manufacturing, assembly, and inspection. A promising use of robots is in surroundings that are hazardous to humans. For example, robots can be used to perform tasks in the presence of hazardous substances that would threaten human health and safety.

Computers

The explosive growth of digital computer hardware and software is one of the most interesting and arguably the most influential phenomena of our time. Computers have found applications throughout manufacturing. The most important of these are outlined here.

Computer-aided design (CAD) is employed to convert an idea to a manufactured product. Whereas innumerable sketches, engineering drawings, and physical mock-ups used to be required to bring this transition about, computer graphics are now used. Thus computers can be used to provide a realistic visual picture of a product, to analyze its characteristics and performance, and to redesign it based upon the results of computer analysis. The capabilities of computers in this respect are enormous. As an example, Boeing's large extremely complex 770 passenger airliner, which entered commercial service in 1995, was designed by computer and brought to production without construction of a full-scale mockup.

Closely linked to CAD is *computer-aided manufacturing*, CAM which employs computers to plan and control manufacturing operations, for quality control, and to manage entire manufacturing plants. The CAD/CAM combination continues to totally change manufacturing operations to the extent that it may be called a "new industrial revolution."

The application of computers has had a profound influence on environmental concerns. One example is the improved accuracy of weather forecasting that has resulted form sophisticated and powerful computer programs and hardware. Related to this are the uses of weather satellites, which could not be placed in orbit or operated without computers. Satellites operated by computer control are used to monitor pollutants and map their patterns of dispersion. Computers are widely used in modelling to mimic complex ecosystems, climate, and other environmentally relevant systems.

Computers and their networks are susceptible to mischief and sabotage by outsiders. The exploits of "*computer hackers*" in breaking into government and private sector computers have been well documented. Important information has been stolen and the operation of computers has been seriously disrupted by hackers with malicious intent. Most computer operations are connected with others through the internet, enabling communication with employees at remote locations and instant contact with suppliers and customers. The problem of deliberate disruption is potentially so great that in 1997 companies throughout the world spend about $6 billion on outside experts in computer security, a figure that is expected to double around the year 2000. The U.S. Federal Bureau of Investigation (FBI) now trains its new agents in cyberspace crime, and maintains special computer crime squads in New York, Washington, and San Francisco.

Numerous kinds of protection are available for computer installations. Such protection comes in the form of both software and hardware. Special encryption software can be used to put computer messages in code that is hard to break. Hardware and software barriers to unauthorized corporate computer access, "firewalls," continue to become more sophisticated and effective.

HUMAN IMPACT AND ENVIRONMENTAL POLLUTION

The challenge facing modern, technologically based societies is to achieve and maintain a high standard of living and quality of life without ruining and indeed while enhancing the Earth support system upon which society depends for its existence. Ultimately, this has to be done on a global basis, although much can be done nationally, locally, and individually. In this respect, the education and training of people who will be directing the technology on which society operates is of utmost importance. Traditional, narrowly based education in areas such as chemistry, economics, engineering, and even ecology are not suitable to prepare people for this challenge. The development of a "booming economy" defined in a traditional sense and characterized by high rates of production, consumption, and development can be very bad for the environment and is ultimately unsustainable. Engineering solutions that do not consider environmental protection are similarly undesirable. A purely environmental solution and a "back-to-the-land" approach can result in great hardship and are unacceptable to the majority of people. What is needed, then, are systems that interweave all the aspects of a modern industrialized society with an environmentally enlightened approach.

The term *sustainable development* has been used to describe industrial development that can be sustained without environmental damage and to the benefit of all people. A "post-industrial world" built around sustainable societies has been outlined. Although sustainable development has become widely used, it has been pointed out that some consider the term to be "an oxymoron without substance." Clearly, if humankind is to survive with a reasonable standard of living, something like "sustainable development" must evolve in which use of nonrenewable resources is minimized insofar as possible, and the capability to produce renewable resources (for example, by promoting soil conservation to maintain the capacity to grow biomass) is enhanced. This will require significant behavioral changes, particularly in limiting population growth and curbing humankind's appetite for increasing consumption of goods and energy.

3

ATMOSPHERIC CHEMISTRY

The quality of the air that living organisms breathe, the nature and level of air pollutants, visibility and atmospheric esthetics, and even climate are dependent upon chemical phenomena that occur in the atmosphere. These, in turn are strongly tied to absorption of solar energy, interactions between the gas phase of the atmosphere and small solid particles suspended in it, and interchange of chemical species with the geosphere. Chemical reactions in the atmosphere and the chemical nature of atmospheric chemical species are the topic of *atmospheric chemistry*, which is introduced in this chapter.

Atmospheric chemistry involves the unpolluted atmosphere, highly polluted atmospheres, and a wide range of gradations in between. The same general phenomena govern all the produce one huge atmospheric cycle, in which there are numerous subcycles. Gaseous atmospheric chemical species fall into the following somewhat arbitrary and overlapping classifications: inorganic oxides (CO, CO_2, NO_2, SO_2); oxidants (O_3); reductants (CO, SO_2, H_2S); organics (in the unpolluted atmosphere, CH_4 is the predominant organic species, whereas alkanes, alkenes, and aryl compounds are common around sources of organic pollution); photochemically active species (NO_2, formaldehyde); acids (H_2SO_4); bases (NH_3); salts (NH_4HSO_4); and unstable reactive species (electronically excited NO_2, HO• radical). In addition, both solid and liquid particles play a strong role in atmospheric chemistry as sources and sinks for gas-phase species, as sites for surface reactions (solid particles), and as bodies for aqueous-phase reactions (liquid droplets). Two constituents of utmost importance in atmospheric chemistry are radiant energy from the sun, predominantly in the ultraviolet region of

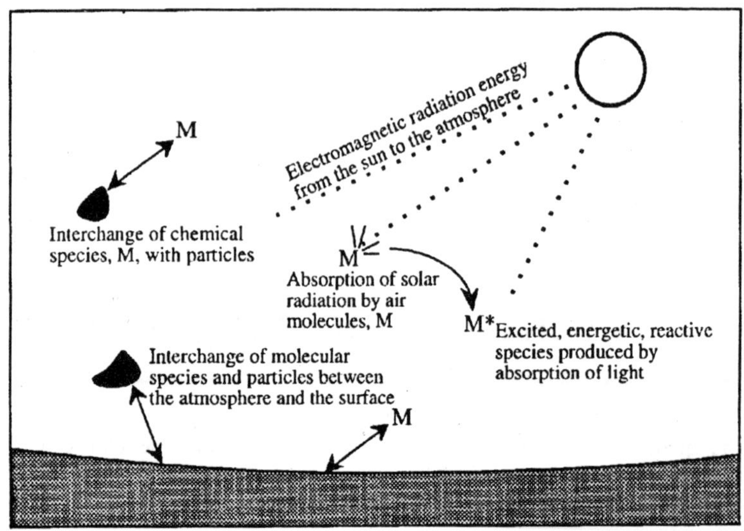

Fig. 3.1. Illustration of atmospheric chemical processes.

the spectrum, and the hydroxyl radical, HO•. The forme provides a way to pump a high level of energy into a single gas molecule to start a series of atmospheric chemical reactions, and the latter is the most important reactive intermediate and "currency" of daytime atmospheric chemical phenomena. NO_3 radicals are important intermediates in nighttime atmospheric chemistry.

It is difficult to study atmospheric chemical reactions in the laboratory. In addition to the very low concentrations of species involved, walls required to contain gases in the laboratory evolve contaminant species, serve as third bodies to absorb energy from atmospheric chemical reactions, and act as catalysts. These effects are particularly pronounced when attempts are made to replicate the low-pressure conditions of the upper atmosphere.

Photochemical Processes

The absorption of electromagnetic solar radiation (light) by chemical species causes *photochemical reactions*. Photochemical reactions give atmospheric chemistry a unique quality and largely determine the nature and ultimate fate of atmospheric chemical species. The ability of electromagnetic radiation to cause photochemical reactions to occur is a function of its energy, E, which increases with increasing frequency (ν) and decreasing wavelength (λ) according to the relationship.

$$E = h\nu \qquad \ldots(1)$$

where h is Planck's constant, 6.62×10^{-27} erg sec. Radiation capable of causing photochemical reactions is called *actinic radiation*. In order for a photochemical reaction to occur, a single unit of photochemical energy form actinic radiation, called a *quantum* and having an energy of $h\nu$, must be absorbed by the reacting species. If the absorbed light is in the visible region of the sun's spectrum, the absorbing species is colored. Colored NO_2 is a common example of such a species in the atmosphere.

An important characteristic of photochemical reactions is the degree to which a particular and result occurs form the absorption of photons. This is expressed by the *quantum yield*. For the hypothetical case in which product "B" is produced as the result of absorption of photons by reactant "A," the quantum yield is the following:

$$\text{Quantum yield} = \frac{\text{Number of molecules of B}}{\text{Number of photons absorbed by A}} \quad ...(2)$$

Nitrogen dioxide, NO_2, is one of the most photochemically active species found in a polluted atmosphere. When a molecule such as NO_2 absorbs actinic radiation of energy $h\nu$,

$$NO_2 + h\nu \rightarrow NO_2{}^* \quad ...(3)$$

an *electronically excited molecule* designated in the reaction above by an asterisk,*, ay be produced.

Electronically excited molecules and atoms are reactive and unstable species in the atmosphere that participate in a wide range of atmospheric chemical processes. Two other generally reactive and unstable species in the atmosphere are *free radicals* composed of atoms or molecular fragments with unshared electrons, and *ions* consisting of charged atoms or molecular fragments.

Electronically excited molecules produced when unexcited ground-state molecules absorb energetic electromagnetic radiation in the ultraviolet or visible regions of the spectrum may possess several possible excited states. Generally, however, ultraviolet or visible radiation is energetic enough to excite molecules only several of the lowest energy levels. The nature of the excited state may be understood by considering the disposition of electrons in a molecule. Most molecules have an even number of electrons. Electrons in molecules and atoms occupy orbitals, with a maximum of two electrons with opposite spin in the same orbital. The absorption of light may promote one of these electrons to a vacant higher-energy orbital. In some cases the promoted electron retains a spin opposite to that of its forme partner, giving rise to an *excited singlet state*. In other cases the spin of the promoted electron is reversed, such that it has the same spin as its former partner; this gives rise to an *excited triplet state*.

These excited states are relatively energized compared to the ground state; therefore, excited chemical species are reactive. Their participation in atmospheric chemical reactions, such as those involved in smog formation, are discussed later in detail.

Electromagnetic radiation absorbed in the infrared region is not sufficiently energetic to break chemical bonds, but does cause the receptor molecules to gain vibrational and rotational energy. The energy absorbed as infrared radiation ultimately is dissipated as hear and raises the temperature of the whole atmosphere.

The reactions that occur following absorption of a photon of light sufficiently energetic to produce an electronically excited species are largely determined by the way in which the excited species loses its excess energy. This may occur by one of several processes divided into two general classes. Of these, *photophysical processes* are those that do not involve chemical bond breakage or loss of electrons and include loss of energy from the excited molecule by electromagnetic radiation as it returns to the ground state (fluorescence or phosphorescence), transfer of energy to other molecules, or transfer of energy within the absorbing molecule; the last two processes result in dissipation of the excess energy as heat. Photochemical reactions occur as a result of de-excitation processes that involves chemical bond breakage or ion formation, particularly the following:

1. *Photodissociation* of the excited molecule (the process responsible for the predominance of atomic oxygen in the upper atmosphere)

$$O_2^* \rightarrow O + O \qquad ...(4)$$

2. *Direct reaction* with another species

$$O_2^* + O_3 \rightarrow 2O_2 + O \qquad ...(5)$$

3. *Photoionization* through loss of an electron

$$N_2^* \rightarrow N_2^+ + e^- \qquad ...(6)$$

Photochemically excited molecules may undergo processes other than those listed above. One of these is *intramolecular rearrangement*, such as occurs when 2-nitrobenzaldehyde absorbs light:

$$C_6H_4(CHO)(NO_2) + h\nu \rightarrow C_6H_4(COOH)(NO)$$

This reaction occurs so readily that it is used to measure light intensity. *Photoisomerization* can also occur, such as shown by the example below in which the H atom and CH_3 group switch positions when the compound absorbs a photon:

Absorption of *hν* causes isomerization as H_3C and H switch positions.

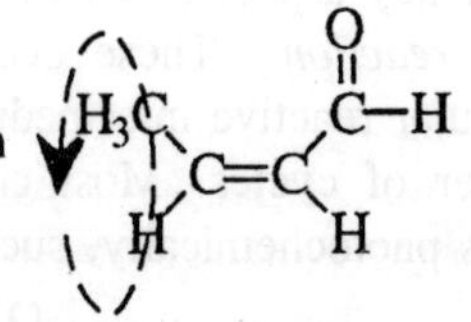

$$\begin{array}{c} H_3C \qquad \overset{O}{\overset{\|}{C}}-H \\ \quad C=C \\ H \qquad\quad H \end{array}$$

Photochemically excited molecules may also under *dimerization*, in which two molecules join together.

Insofar, as gas-phase reactions in the troposphere are concerned, the most important of the processes listed above is photodissociation. This is because photodissociation converts relatively stable and unreactive molecular species to reactive atoms and free radicals that participate in additional reactions, including chain reactions.

CHEMICAL PROCESSES AND CHAIN REACTIONS IN THE ATMOSPHERE

As noted in the preceding section, energetic electromagnetic radiation in the atmosphere may produce atoms or groups of atoms with unpaired electrons called free radicals:

$$H_3C-\overset{\overset{O}{\|}}{C}-H + h\nu \rightarrow H_3C\bullet + H\dot{C}O \qquad ...(7)$$

In the formula above, the single dot, •, represents the unpaired electron that makes free radicals so chemically reactive. Free radicals are involved with most significant atmospheric chemical phenomena and are of the utmost importance in the atmosphere. Because of their unpaired electrons and are strong pairing tendencies of electrons under most circumstances, free radicals are highly reactive; therefore, they generally have short lifetimes. It is important to distinguish between high reactivity and instability. A totally isolated free radical or a single atom, such as an O atom, would be quite stable; it wants to react, but there is nothing around for it to react with. Therefore, free radicals and single atoms from diatomic gases (such as O from O_2) tend to persist under the rarefied conditions of very high altitudes because they can travel long distances before colliding with another reactive species. However, unlike free radicals, electronically excited species have a finite, generally very short, lifetime because they can lose energy through radiation without having to react with another species.

Chain Reactions

A key aspect of chemical processes in the atmosphere is that of *chain reactions*. These occur when a series of reactions involving particular reactive intermediates, usually free radicals, goes through a number of cycles. Most commonly, an atmospheric chain reaction begins photochemically, such as by the photodissociation of NO_2:

$$NO_2 + h\nu \rightarrow NO + O \quad ...(8)$$

Radicals can take part in chain reactions in which one of the products of each reaction is a radical. Eventually, through processes such as reaction with another radical, one of the radicals in a chain is destroyed and the chain ends:

$$H_3C^{\bullet} + H_3C^{\bullet} \rightarrow C_2H_6 \quad ...(9)$$

This process is a *chain-terminating reaction*. Reactions involving free radicals are responsible for smog formation.

Hydroxyl and Hydroperoxyl Radicals in the Atmosphere

The hydroxyl radial, $HO^{\bullet}$, which is the single most important reactive intermediate species in atmospheric chemical processes, is formed by several mechanisms. At higher altitudes it is produced by photolysis of water:

$$H_2O + h\nu \rightarrow HO^{\bullet} + H \quad ...(10)$$

In the presence of organic matter, hydroxyl radical is produced in abundant quantities as an intermediate in the formation of photochemical smog. To a certain extent in the atmosphere, and for laboratory experimentation, $HO^{\bullet}$ is made by the photolysis of nitrous acid vapor:

$$HONO + h\nu \rightarrow HO^{\bullet} + NO$$

In the relatively unpolluted troposphere, hydroxyl radial is produced as the result of the photolysis of ozone, followed by the reaction of a fraction of the excited oxygen atoms with water molecules:

$$O_3 + h\nu(\lambda < 315 \text{ nm}) \rightarrow O^* + O_2 \quad ...(11)$$

$$O^* + H_2O \rightarrow 2HO^{\bullet} \quad ...(12)$$

Among the important atmospheric trace species that react with hydroxyl radical and are thus removed from the atmosphere are carbon monoxide, sulfur dioxide, hydrogen sulfide, methane, and nitric oxide.

Hydroxyl radical is most frequently removed from the troposphere by reaction with methane or carbon monoxide:

$$CH_4 + HO^{\bullet} \rightarrow H_3C^{\bullet} + H_2O \quad ...(13)$$

$$CO + HO^{\bullet} \rightarrow CO_2 + H \qquad ...(14)$$

The reactive methyl radical, $H_3C^{\bullet}$, and the hydrogen atom produced in the preceding reactions undergo additional reactions in the atmosphere. The global concentration of hydroxyl radical, averaged diurnally and seasonally, is estimated to range from 2×10^5 to 1×10^6 radicals per cm^3 in the troposphere. Because of the greater humidity and higher incident sunlight, which result in elevated O^* levels, the concentration of $HO^{\bullet}$ is relatively higher in tropical regions. The Southern Hemisphere probably has about a 20% higher level of $HO^{\bullet}$ than does the Northern Hemisphere because there is more $HO^{\bullet}$-consuming CO produced by human activities in the Northern Hemisphere.

Oxidation Processes in the Atmosphere

The 21 percent (dry basis) by volume content of molecular O_2 makes the atmosphere thermodynamically oxidizing. One prominent manifestation of this condition is the tendency for oxidizable materials to corrode when exposed to the atmosphere. Iron, for example, exposed to moist air tends to rust:

$$4Fe + 3O_2 + xH_2O \rightarrow 2Fe_2O_3{\cdot}xH_2O \qquad ...(15)$$

From the standpoint of atmospheric chemistry, however, the oxidizing tendency of the atmosphere is shown by the conversion of reduced molecular species to oxidized forms. It is this feature of the atmosphere exposed to sunlight that results in the formation of photochemical smog. Among the simple molecular species that enter the atmosphere in relatively reduced forms and that are oxidized are the following:

$$2CO + O_2 \rightarrow 2CO_2 \qquad ...(16)$$

$$CH_4 + 2O_2 \rightarrow 2CO_2 + 2H_2O \qquad ...(17)$$

$$4NO + 3O_2 + 2H_2O \rightarrow 4HNO_3 \qquad ...(18)$$

$$2SO_2 + O_2 + 2H_2O \rightarrow 2H_2SO_4 \qquad ...(19)$$

$$H_2S + 2O_2 \rightarrow H_2SO_4 \qquad ...(20)$$

Although shown here in a very simple form, these reaction actually represent processes that may involve many steps, photochemistry, and reactive intermediates, particularly hydroxyl radical. Oxidation reactions may also occur on particle surfaces and in solution in aqueous aerosol droplets, which are strongly exposed to atmospheric oxygen. Another aspect of these reactions is that the products are often acidic—mildly acidic CO_2 from carbon-containing species, and strongly acidic nitric and sulfuric acid from nitrogen oxides and gasecus sulfur species, respectively.

Reducing agents, such as those shown above, may be quite stable in dry air that is not exposed to sunlight. However, the absorption of photons from solar radiation starts processes that result in oxidation. As a simple example, the photochemical dissociation of nitrogen dioxide,

$$NO_2 + h\nu \rightarrow NO + O \qquad ...(21)$$

can produce reactive O atoms that can react with oxidizable molecules.

$$CH_4 + O \rightarrow H_3C^{\bullet} + HO^{\bullet} \qquad ...(22)$$

to begin the series of reactions that forms the final oxidized products (in this case CO_2 and H_2O). Intermediate hydroxyl radical, $HO^{\bullet}$, can abstract H atoms from hydrocarbons,

$$CH_4 + HO \rightarrow H_3C^{\bullet} + H_2O \qquad ...(23)$$

or add to molecules such as NO_2,

$$HO^{\bullet} + NO_2 \rightarrow HNO_3 \qquad ...(24)$$

to bring about oxidations. Chain reactions can be involved, such as following sequence that regenerates NO_2 from NO:

$$H_3C\bullet\text{(from Reactions 22)} + O_2 \rightarrow H_3COO\bullet \qquad ...(25)$$

$$H_3COO\bullet + NO \rightarrow NO_2 \text{ (back to Reaction 21)} + H_3CO\bullet \qquad ...(26)$$

The NO_2 product may undergo photochemical dissociation to produce O atoms and again initiate processes that result in oxidation.

A feature of the photochemical atmosphere, particularly when it is polluted by nitrogen oxides and hydrocarbons, is the generation of strong oxidant molecules. The most common example of a strong organic oxidant species is peroxyacetyl nitrate, PAN, formed from the reaction of $H_3CC(O)OO^{\bullet}$ radical with NO_2:

$$H_3CC(O)OO^{\bullet} + NO_2 + \text{M(energy-absorbing third molecule)} \rightarrow$$
$$CH_3C(O)OONO_2 + M \qquad ...(27)$$

The most prominent inorganic oxidant is ozone, generated by reactions such as,

$$O + O_2 + \text{M (energy-absorbing third molecule)} \rightarrow O_3 + M \qquad ...(28)$$

One of the ways in which ozone acts as an oxidant is to add to unsaturated compounds to form reactive ozonides:

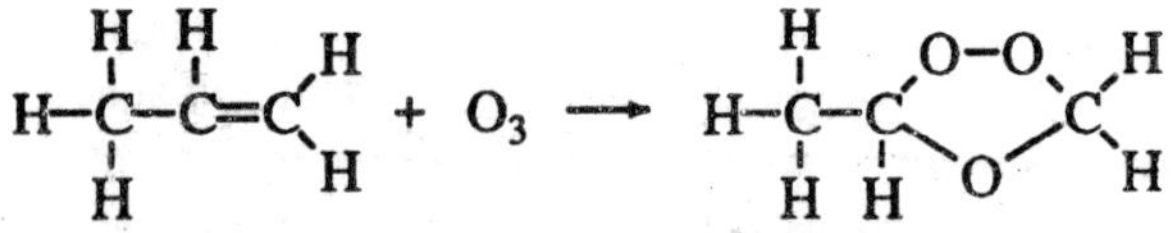

```
  H  H                        H
  |  |     H                  |    O—O    H
H—C—C=C          + O3  →   H—C—C      C
  |       H                   |  |  \  /   H
  H                           H  H   O
```

ACID-BASE REACTIONS IN THE ATMOSPHERE

Acid-base reactions occur between acidic and basic species in the atmosphere. The atmosphere is normally at least slightly acidic because of the presence of a low level of carbon dioxide, which dissolves in atmospheric water droplets and dissociates slightly:

$$CO_2(g) \xrightarrow{\text{Water}} CO_2(aq) \qquad ...(29)$$

$$CO_2(aq) + H_2O \rightarrow H^+ + HCO_3^- \qquad ...(30)$$

Atmospheric sulfur dioxide forms a somewhat stronger acid when it dissolves in water:

$$SO_2(g) + H_2O \rightarrow H^+ + HSO_3^- \qquad ...(31)$$

In terms of pollution, however, strongly acidic HNO_3 and H_2SO_4 formed by the atmospheric oxidation of N oxides, SO_2, and H_2S are much more important because they lead to the formation of damaging acid rain.

As reflected by the generally acidic pH of rainwater, basic species are relatively less common in the atmosphere. Particulate calcium oxide, hydroxide, and carbonate can get into the atmosphere from ash and ground rock and can react with acids such as in the following reaction:

$$Ca(OH)_2(s) + H_2SO_4(aq) \rightarrow CaSO_4(s) + H_2O \qquad ...(32)$$

The most important basic species in the atmosphere is gas-phase ammonia, NH_3. The greatest source of atmospheric ammonia is from biodegradation of nitrogen-containing biological matter and from bacterial reduction of nitrate:

$$NO_3^-(aq) + 2\{CH_2O\}(biomass) + H^+ \rightarrow NH_3(g) + 2CO_2 + H_2O \qquad ...(33)$$

Ammonia is particularly important as a base in the atmosphere because it is the only water-soluble base present at significant levels in the atmosphere. Dissolved in atmospheric water droplets, it plays a strong role in neutralizing atmospheric acids:

$$NH_3(aq) + HNO_3(aq) \rightarrow NH_4NO_3(aq) \qquad ...(34)$$

$$NH_3(aq) + H_2SO_4(aq) \rightarrow NH_4HSO_4(aq) \qquad ...(35)$$

These reactions have three effects: (1) They result in the presence of NH_4^+ ion in the atmosphere as dissolved or solid salts, (2) they serve in part to neutralize acidic constituents of the atmosphere, and (3) they produce relatively corrosive ammonium salts.

Ions in the Atmosphere

One of the characteristics of the upper atmosphere above the troposphere which is difficult to duplicate under laboratory conditions is the presence of significant levels of electrons and positive ions. Because of the rarefied conditions, these ions may exist in the upper atmosphere for long periods before recombining to form neutral species.

At altitudes of approximately 50 km and up, ions are so prevalent that the region is called the *ionosphere*. The presence of the ionosphere has been known since about 1901, when it was discovered that radio waves could be transmitted over long distances, where the curvature of the Earth makes line-of-sight transmission impossible. These radio waves bounce off the ionosphere.

Ultraviolet radiation is the primary producer of ions in the ionosphere:

$$N_2 + h\nu \rightarrow N_2^+ + e^- \qquad ...(36)$$

Other ions are produced by secondary reactions, such as the following:

$$N_2^+ + O \rightarrow NO^+ + N \qquad ...(37)$$

In darkness, the positive ions produced by photochemical processes slowly recombine with free electrons. The process is especially rapid in the lower regions of the ionosphere, where the concentration of species is relatively high. Thus, the lower boundary of the ionosphere lifts at night and makes possible the transmission of radio waves over much greater distances.

The Earth's magnetic field has a strong influence upon the ions in the upper atmosphere. Probably the best-known manifestation of this phenomenon is found in the Van Allen belts, discovered in 1958. These regions consist of two belts of ionized particles circling the Earth. If they are visualized as two doughnuts, then the axis of the Earth's magnetic field extends through the holes in the doughnuts. In the inner

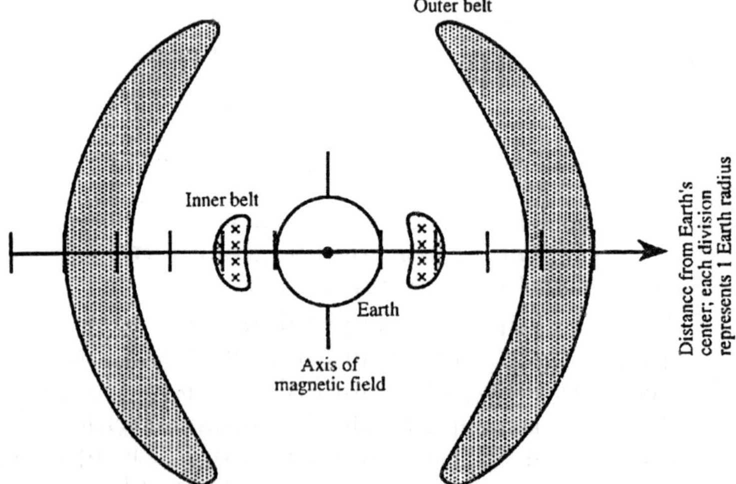

Fig. 3.2. Cross section of the Van Allen belts encircling the Earth.

belt, the highly energetic ionizing radiation consists of protons. In the outer belt, it consists of electrons.

Ions, produced in the upper atmosphere primarily by the action of energetic electromagnetic radiation, may also be produced in the troposphere by the shearing of water droplets during precipitation. The shearing may be caused by the compression of descending masses of cold air or by strong winds over hot, dry land masses. The last phenomenon is known as the foehn, sharav (in the Near East), or Santa Ana (in southern California). These hot, dry winds cause sever discomfort. The ions produced by them consist of electrons and positively charged molecular species. With sufficient humidity, these ions quickly become surrounded by clusters of one to eight water molecules forming what are called small air ions that may last for up to several minutes. Small air ions undergo further reactions, including incorporation of trace gases from the atmosphere. Eventually, they are neutralized by combining with ions of the opposite charge or with uncharged condensation particles.

Evolution of the Atmosphere

Earth's atmosphere originally was very different form its present state and has changed largely through biological activity and accompanying chemical changes. According to theories about the origin of life that have been widely accepted, at least until recently, approximately 3.5 billion years ago, when the first primitive life molecules were formed, the atmosphere was chemically reducing, consisting primarily of methane, ammonia, water vapor, and hydrogen. Bombardment by intense, bond-breaking ultraviolet light, along with lightning and radiation from radionuclides, provided the energy to bring about chemical reactions that resulted in the production of relatively complicated molecules, including even amino acids and sugars. Thus a rich chemical mixture was formed in the sea from which life molecules evolved. Initially, these very primitive life forms derived their energy from fermentation of organic matter formed by chemical and photochemical processes, then gained the ability to produce organic matter, "$\{CH_2O\}$," by photosynthesis,

$$CO_2 + H_2O + h\nu \rightarrow \{CH_2O\} + O_2(g) \quad ...(38)$$

and the stage was set for the massive biochemical transformation that resulted in the production of almost all the atmosphere's oxygen.

The oxygen initially produced by photosynthesis was probably quite toxic to primitive life forms. However, much of this oxygen was converted to iron oxides by reaction with soluble iron(II):

$$4Fe^{2+} + O_2 + 4H_2O \rightarrow 2Fe_2O_3 + 8H^+ \quad ...(39)$$

This process formed enormous deposits of iron oxides, the existence of which provides convincing evidence for the liberation of free O_2 in the primitive atmosphere.

Eventually, enzyme systems developed that enabled organisms to mediate the reaction of waste-product oxygen with oxidizable organic matter in the sea. Later, this mode of waste-product disposal was utilized by organisms to produce energy by respiration, which is now the mechanism by which nonphotosynthetic organisms obtain energy. In time, O_2 accumulated in the atmosphere. In addition to providing an abundant source of oxygen for respiration, the accumulated atmospheric oxygen formed an ozone shield. The ozone shield absorbs bond-rupturing ultraviolet radiation. With the ozone shield protecting tissue from destruction by high-energy ultraviolet radiation, the Earth became a much more hospitable environment for life, and life forms were enabled to move from the sea to land.

In the late 1990s an alternate to the above theory for the origin of life gained credence. This occurred because of discoveries of abundant life forms at very great depths around hydrothermal vents on the sea floor. Some authorities contend that these conditions, hostile though they are to more familiar life forms, were those under which the self-replicating molecules required for life began.

Reactions of Atmospheric Oxygen

The oxygen cycle is critically important in atmospheric chemistry, geochemical transformations, and life processes.

Oxygen in the troposphere plays a strong role in processes that occur on the Earth's surface. Atmospheric oxygen takes part in energy-producing reactions, such as the burning of fossil fuels:

$$CH_4(\text{in natural gas}) + 2O_2 \rightarrow CO_2 + 2H_2O \quad ...(40)$$

Atmospheric oxygen is utilized by aerobic organisms in the degradation of organic material. Some oxidative weathering processes consume oxygen. An important example is the oxidative weathering of mineral iron(II) to iron(III):

$$4FeO + O_2 \rightarrow 2Fe_2O_3 \quad ...(41)$$

Oxygen is returned to the atmosphere through plant photosynthesis:

$$CO_2 + H_2O + h\nu \rightarrow \{CH_2O\} + O_2 \quad ...(42)$$

All molecular oxygen now in the atmosphere is thought to have originated through the action of photosynthetic organisms, which shows the importance of photosynthesis in the oxygen balance of the

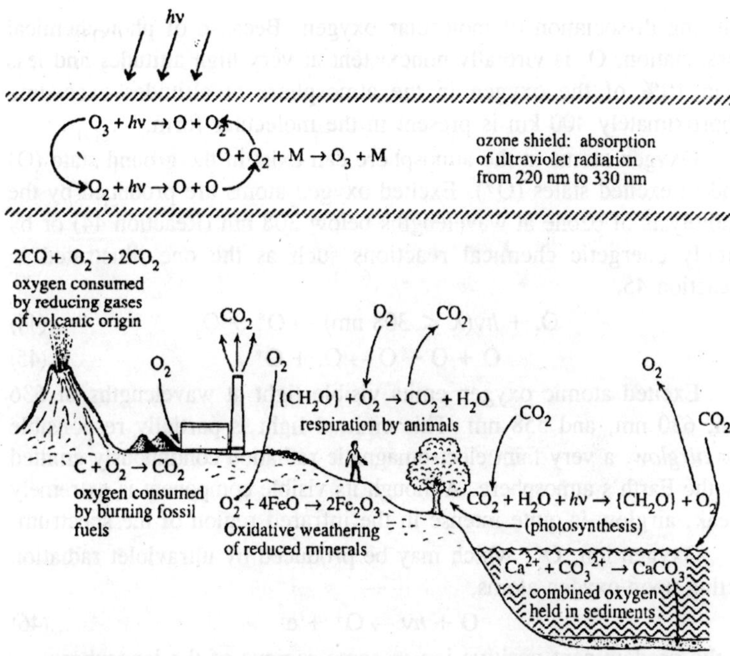

Fig. 3.3. Oxygen exchange among the atmosphere, lithosphere, hydrosphere, and biosphere.

atmosphere. It can be shown that most of the carbon fixed by these photosynthetic processes is dispersed in mineral formations as humic materials; only a very small fraction is deposited in fossil fuel beds. Therefore, although fossil fuel combustion consumes large amounts of O_2, there is not danger of running out of atmospheric oxygen.

Because of the extremely rarefied atmosphere and the effects of ionizing radiation, elemental oxygen in the upper atmosphere exists to a large extent in forms other than diatomic O_2. In addition to O_2, the upper atmosphere contains oxygen atoms, O; excited oxygen molecules, O_2*; and ozone, O_3.

Atomic oxygen, O, is stable primarily in the thermosphere, where the atmosphere is so rarefied that the three-body collisions necessary for the chemical reaction of atomic oxygen seldom occur (the third body in this kind of three-body reaction absorbs energy to stabilize the products). Atomic oxygen is produced by a photochemical reaction:

$$O_2 + h\nu \rightarrow O + O \qquad \text{...(43)}$$

The oxygen-oxygen bond is strong and ultraviolet radiation in the wavelength regions 135-176 nm and 240-260 nm is most effective in causing dissociation of molecular oxygen. Because of photochemical dissociation, O_2 is virtually nonexistent at very high altitudes and less than 10% of the oxygen in the atmosphere at altitudes exceeding approximately 400 km is present in the molecular form.

Oxygen atoms in the atmosphere can exist in the ground state (O) and in excited states (O*). Excited oxygen atoms are produced by the photolysis of ozone at wavelengths below 308 nm (Reaction 44) or by highly energetic chemical reactions such as the one illustrated in Reaction 45.

$$O_3 + h\nu(\lambda < 308 \text{ nm}) \rightarrow O^* + O_2 \quad ...(44)$$

$$O + O + O \rightarrow O_2 + O^* \quad ...(45)$$

Excited atomic oxygen emits visible light at wavelengths of 636 nm, 630 nm, and 558 nm. This emitted light is partially responsible for *airglow*, a very faint electromagnetic radiation continuously emitted by the Earth's atmosphere. Although its visible component is extremely weak, airglow is quite intense in the infrared region of the spectrum.

Oxygen ion, O^+, which may be produced by ultraviolet radiation acting upon oxygen atoms,

$$O + h\nu \rightarrow O^+ + e^- \quad ...(46)$$

is the predominant positive ion in some regions of the ionosphere.

Ozone and the Ozone Layer

Ozone, O_3, has an essential protective function because it absorbs harmful ultraviolet radiation in the stratosphere and serves as a radiation shield, protecting living beings on the Earth from the effects of excessive amounts of such radiation. It is produced by a photochemical reaction,

$$O_2 + h\nu \text{ (energetic ultraviolet radiation)} \rightarrow O + O \quad ...(47)$$

followed by a three-body reaction,

$$O + O_2 + M \rightarrow O_3 + M\text{(increased energy)} \quad ...(48)$$

in which M is another molecule, usually N_2 or O_2, which absorbs the excess energy given off by the reaction and enables the ozone molecule to stay together. The region of maximum ozone concentration occurs in the stratosphere at an altitude of 2-30 km where it may reach levels of 10 ppm.

Ozone absorbs ultraviolet light very strongly in the region 220-330 nm. If this light were not absorbed by ozone, severe damage would result to exposed forms of life on the Earth. Absorption of electromagnetic radiation by ozone converts the radiation's energy to heat and is responsible for the temperature maximum encountered at the boundary between the stratosphere and the mesosphere at an altitude of approximately 50 km. The reason that the temperature maximum occurs at a higher altitude than that of the maximum ozone concentrations arises from the fact that ozone is such an effective absorber of ultraviolet light. Therefore, most of this radiation is absorbed in the upper stratosphere, where it generates heat, and only a small fraction reaches the lower altitudes, which remain relatively cool.

Thermodynamically, the overall reaction,

$$2O_3 \rightarrow 3O_2 \qquad \text{...(49)}$$

is favored so that ozone is inherently unstable. Its decomposition in the stratosphere is catalyzed by a number of natural and pollutant trace constituents, including NO, NO_2, H, $HO^\bullet$, $HOO^\bullet$, ClO, Cl, Br, and BrO. Ozone decomposition also occurs on solid surfaces, such as metal oxides and salts produced by rocket exhausts.

Ozone is an undesirable pollutant in the troposphere. It is toxic, and a mild overdose causes labored breathing, a feeling of chest pressure, cough, and irritated eyes.

Reactions of Atmospheric Nitrogen

The 78% by volume of nitrogen contained in the atmosphere constitutes an inexhaustible reservoir of that essential element. A small amount of nitrogen is thought to be fixed (chemically bound to other elements) in the atmosphere by lighting, and some is also fixed by combustion processes, as in the internal combustion engine.

Before the use of synthetic fertilizers reached its current high levels, chemists were concerned that denitrification processes in the soil would lead to nitrogen depletion on the Earth. Now, with millions of tons of synthetically fixed nitrogen being added to the soil each year, concern has shifted to possible excess accumulation of nitrogen in soil, fresh water, and the oceans.

Unlike oxygen, which is almost completely dissociated to the monatomic form in higher regions of the thermosphere, molecular nitrogen is not readily dissociated by ultraviolet radiation. However, at altitudes exceeding approximately 100 km, atomic nitrogen is produced by photochemical reactions:

$$N_2 + h\nu \rightarrow N + N \qquad \text{...(50)}$$

Most stratospheric ozone is probably removed by the actin of nitric oxide, which reacts with ozone as follows:

$$O_3 + NO \rightarrow NO_2 + O_2 \qquad \text{...(51)}$$

$$NO_2 + O \rightarrow NO + O_2 \text{ (regeneration of NO from } NO_2) \qquad \text{...(52)}$$

Pollutant oxides of nitrogen, particularly NO_2, are key species involved in air pollution and formation of photochemical smog. For example, NO_2 is readily dissociated photochemically to NO and reactive atomic oxygen:

$$NO_2 + h\nu \rightarrow NO + O \quad ...(53)$$

This reaction is the most important primary photochemical process involved in smog formation.

ATMOSPHERIC CARBON DIOXIDE

Although only about 0.035% (350 ppm) of air consists of carbon dioxide, it is the atmospheric "nonpollutant" species of most concern. As mentioned above, carbon dioxide, along with water vapor, is primarily responsible for the absorption of infrared energy re-emitted by the Earth such that some of this energy is reradiated back to the Earth's surface. Current evidence suggests that changes in the atmospheric carbon dioxide level will substantially alter the Earth's climate through the greenhouse effect.

Valid measurements of overall atmospheric CO_2 can only be taken in areas remote from industrial activity. Such areas include Antarctica and the top of Mauna Loa Mountain in Hawaii. Measurements of carbon dioxide levels in these locations over the last 40 years suggest an annual increase in CO_2 of about 1 ppm per year.

The most obvious factor contributing to increased atmospheric carbon dioxide in consumption of carbon-containing fossil fuels. In addition, release of CO_2 from the biodegradation of biomass and uptake by photosynthesis are important factors determining overall CO_2 levels in the atmosphere. The role of photosynthesis is illustrated by the seasonal cycle in carbon dioxide levels in the northern hemisphere. Maximum values occur in April and minimum values in late September or October. These oscillations are due to the "photosynthetic pulse," influenced most strongly by forest in middle latitudes. Forests have a much greater influence than other vegetation because in general forest trees carry out more photosynthesis than other kinds of plants, such as prairie grasses. Furthermore, forests store enough fixed, but readily oxidized carbon in the form of wood and humus to have a marked influence on atmospheric CO_2 content. Thus, during the summer months, forests carry out enough photosynthesis to reduce the atmospheric carbon dioxide content markedly. During the winter, metabolism of biota, such as bacteria decay of humus, release a significant amount of CO_2. Therefore, the current worldwide trend toward destruction of forests and conversion of forest lands to agricultural uses will contribute substantially to a grater overall increase in atmospheric CO_2 levels.

With the current trends, it is likely that global CO_2 levels will double by the middle of the next century, which may well raise the Earth's mean surface temperature by 1.5–4.5°C. Such a change might have more potential to cause massive irreversible environmental changes than any other disaster short of global nuclear war.

Chemically and photochemically, CO_2 is a comparatively insignificant species because of its relatively low concentrations and low photochemical reactivity. The infrared radiation absorbed by carbon dioxide is not energetic enough to cause photochemical reactions to occur. The photodissociation of CO_2 by energetic solar ultraviolet radiation is probably a source of CO in the upper atmosphere:

$$CO_2 + h\nu \rightarrow CO + O \qquad \ldots(54)$$

Atmospheric Water

Gaseous water in the upper atmosphere is involved in the formation of hydroxyl and hydroperoxyl radicals as mentioned above. Condensed water vapor in the form of very small droplets is important in atmospheric chemistry. The harmful effects of some air pollutants—for instance, the corrosion of metals by acid-forming gases—require the presence of water, which may come from the atmosphere. Atmospheric water vapor has an important influence upon pollution-induced fog formation under some circumstances. Water vapor interacting with pollutant particulate matter in the atmosphere may

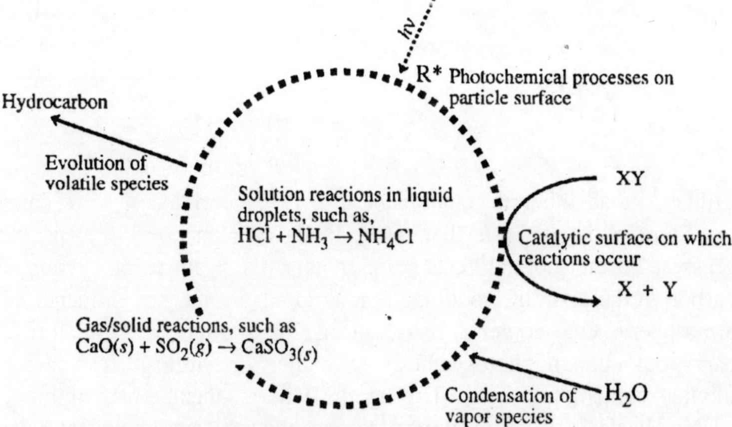

Fig. 3.4. Particles in the atmosphere participate in a number of atmospheric chemical processes.

reduce visibility to undesirable levels through the formation of aerosol particles.

Most stratospheric water comes from the photochemical oxidation of methane:

$$CH_4 + 2O_2 + h\nu \xrightarrow[\text{(several steps)}]{} CO_2 + 2H_2O \qquad ...(55)$$

The water thus produced serves as a source of stratospheric hydroxyl radical as shown by the following reaction:

$$H_2O + h\nu \rightarrow HO^{\bullet} + H \qquad ...(56)$$

ATMOSPHERIC PARTICLES AND ATMOSPHERIC CHEMISTRY

Particles are involved in many chemical reactions in the atmosphere. Neutralization reactions, which occur most readily in solution, may take place in water droplets suspended in the atmosphere. Small particles of metal oxides and carbon have a catalytic effect on oxidation reactions. Particles may also participate in oxidation reactions induced by light.

4

ENVIRONMENTAL CHEMISTRY

Chemistry is the science of matter, where *mater* is anything that has mass and occupies space. The most fundamental kind of mater consists of the *elements*. Of these, *metals*, such as copper or silver, are generally solid, shiny in appearance, electrically conducting, and malleable. *Nonmetals* often have a dull appearance, are not at all malleable, and may exist as gases (atmospheric oxygen), liquids (bromine), or solids (Sulfur). *Organic substances* consist of virtually all compounds, such as wood, that contain carbon. All other substances are *inorganic substances*.

ATOMS, THE BUILDING BLOCKS OF MATTER

As discussed in this chapter, all matter is composed of only about a hundred fundamental kinds of matter called *elements*. Each element is made up of very small entities called *atoms*. Atom, in turn, are composed oft he following very small *subatomic particles*. The two subatomic particles that are located in the small, dense *nucleus* in the center of the atom are the proton, designated *p*, p^+ or +, having a charge of +1 and a mass of 1.007 atomic mass units (u, defined as exactly 1/12 the mass of the most common kind of atom of carbon, carbon-12) and the neutron, *n*, charge 0, mass 1.009 *u*. Moving rapidly around the nucleus of the atom and forming a cloud of negative charge that composes essentially all of the volume of the atom is the electron, designated *e*, e^-, or -, charge -1, mass 0.0005 u. The neutron and proton, each with a mass of essentially 1 u, are said to have a *mass number* of 1, and much lighter electron is said to have a mass number of 0. The charges and mass numbers of the three basic subatomic particles can be denoted by the following three symbols $^{1}_{1}p$, $^{1}_{0}n$ and $^{0}_{-1}e$.

Atoms and Elements

Each atom of a particular element has the same number of protons in its nucleus. This is the *atomic number* of the element. Each element has a *chemical symbol*, such as carbon, C; potassium k (for its Latin name kalium); or cadmium, Cd. Although atoms of the same element are chemically identical, atoms of most elements consist of two or more *isotopes* that have different numbers of neutrons in their nuclei. The *mass number* commonly used to denote isotopes and subatomic particles is the sum of the number of protons and neutrons in the nucleus of the isotope. Each element has an *atomic mass* (atomic weight), the average mass of all atoms of the element, which can be expressed in *atomic mass units*, *u*, also used to express masses of individual atoms, molecules (aggregates of atoms), and subatomic particles.

Important elements

Some elements are more commonly encountered than others. It is useful to be aware of the names, symbols, and properties of some of the most significant elements without having to look them up. Most of the "more important common elements" are among the first 20 elements, which will be discussed next. Several other elements the names and symbols of which should be memorized are copper, Cu, atomic number 29, atomic mass 63.54; iodine, I, atomic number 53, atomic mass 126.904; lead, pb, atomic number 82, atomic mass 207.19; mercury, Hg, atomic number 80, atomic mass 200.59; silver, Ag, atomic number 47, atomic mass 107.87; and zinc, Zn, atomic number 30, atomic mass 65.37.

Elements and the Periodic Table

When elements are considered in order of increasing atomic number, it is observed that their properties are repeated in a periodic manner. For example, elements with atomic numbers, 2, 10, and 18 are gases that do not undergo chemical reactions and consist of individual atoms, whereas those with atomic numbers larger by one—3, 11 and 19—are unstable, highly reactive metals. The *periodic table* is a very useful listing of elements by symbol, atomic number, and atomic mass in a manner that reflects this recurring behavior. The periodic table gets its name from the fact that the properties of elements are repeated periodically in *periods* that go from left to right across a horizontal row of elements. The table is arranged such that an element has properties similar to those of other elements above or below it in the table. Elements with similar chemical properties are called *groups* of elements and are contained in vertical columns in the periodic table. In this section, a periodic table will be developed for the first 20 elements.

Development of the 20-Element Periodic Table

The atom of the first element in the periodic table, *hydrogen*, is the simplest of all, having only one positively charged proton in its nucleus, which is surrounded by a cloud of negative charge formed by only one electron. By far the most abundant kind of hydrogen atom has no neutrons in its nucleus, so its mass number is 1. A small fraction of hydrogen atoms also have 1 neutron (deuterium), and fewer still have 2 neutrons (radioactive tritium). The three different forms of elemental hydrogen are *isotopes* of hydrogen that all have the same number of protons, but different numbers of neutrons in their nuclei; they may be designated as ${}^{1}_{1}H$, ${}^{2}_{1}H$ and ${}^{3}_{1}H$. The average mass of all hydrogen atoms is 1.0079 u (atomic ass units), so hydrogen's *atomic mass* is 1.0079. Hydrogen's box in the periodic table designates its atomic number (1), symbol (H), and atomic mass (1.0079).

Showing electrons in atoms

Electrons in chemical symbols and formulas are shown with *electron-dot symbols* or *Lewis symbols*, which use dots around the symbol of an element to show *outer electrons*. These *valence electrons* are the ones that may become involved in chemical bonds. The hydrogen atom has only one electron, and its Lewis symbol is H•.

Electron configuration and the periodic table

Electron configuration, which determine chemical behavior, describe how electrons in atoms occupy distinct *energy levels* and are contained *electron shells*. Each shell can hold a maximum number of electrons. An atom with a *filled electron shell* has little or no tendency to lose, gain, or share electrons; elements composed of such atoms exist as chemically unreactive gas-phase atoms and are called *noble gases*. The lightest element with filled electron shells is helium, He, atomic number 2. All helium atoms contain 2 protons and 2 electrons. Virtually all helium atoms contain 2 neutrons in their nuclei, and the atomic mass of helium is 4.00260.

The third element in the periodic table is the metal lithium (Li), atomic number 3, atomic mass 6.941. The most abundant lithium isotope has 4 neutrons in its nucleus and a less common isotope has only 3 neutrons. The lithium atom has three electrons. Lithium has both *inner electrons*—in this case 2 contained in an *inner shell*—as in the

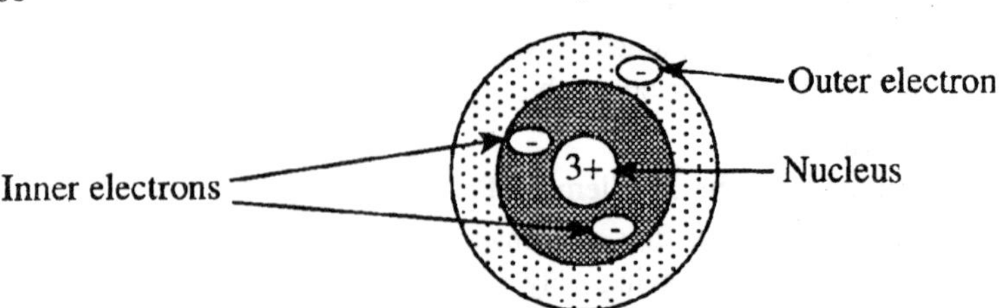

Fig. 4.1. An atom of lithium, Li, has 2 inner electrons and 1 outer electron. The latter can be lost to another atom to produce the Li^+ ion, which is present in ionic compounds.

immediately preceding noble gas helium, and an *outer electron* that is farther from, and less strongly attracted to, the nucleus. The outer electron is said to be in the atom's *outer shell.* The inner electrons are, on the average, closer to the nucleus than is the outer electron, are very difficult to remove from the atom, and do not become involved in chemical bonds. Lithium's outer electron is relatively easy to remove from the atom, which is what happens when ionic bonds involving Li^+ ion are formed.

In atoms such as lithium that have both outer and inner electrons, the Lewis symbol shown only the outer electrons. Therefore, the Lewis symbol of lithium is Li•.

The second period of the table contains elements with atomic number 3 (lithium) through 10. Those other than lithium are the following, listed by atomic number; 4, beryllium, Be; 5 boron, B; 6, carbon, C; 7 nitrogen, N; 8, oxygen, O; 9, fluorine, F; and 10, neon, Ne. The last element in the second period under discussion is *neon.* As shown by its Lewis symbol, :N̈e: the neon atom has 8 outer electrons. These 8 electrons constitute a *filled electron shell*, just as

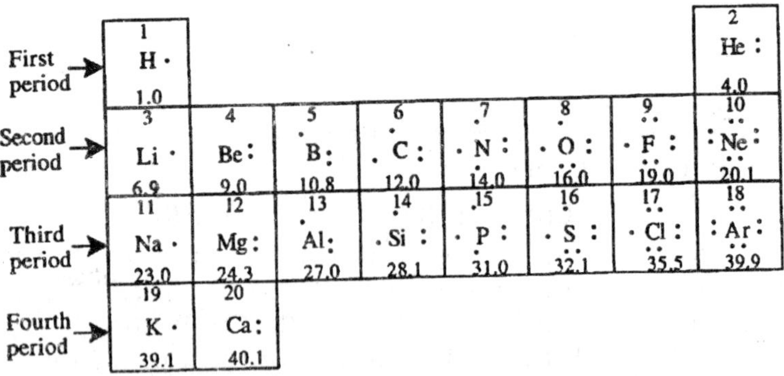

Fig. 4.2. Abbreviated 20 element version of the periodic table showing Lewis symbols of the elements.

the 2 electrons in helium give it a filled electron shell. Because of this "satisfied" outer shell, the neon atom has no tendency to acquire, give away, or share electrons. Therefore, neon is a *noble gas*, like helium, and consist of individual neon atoms

Stability of the neon noble gas electron octet

Like neon, all other atoms with 8 outer electrons are chemically unreactive noble gases. In addition to neon, these are argon (atomic number 18), krypton (atomic number 36), xenon (atomic number 54), and radon (atomic number 86). Each of the noble gases may be represented by the Lewis symbol $:\ddot{\underset{..}{X}}:$, where X is the chemical symbol of the noble gas. It is seen that these atoms each have 8 outer electrons, a group known as an *octet* of electrons.

In many cases atoms that do not have an octet of outer electrons acquire one by losing, gaining, or sharing electrons in chemical combination with other atoms; that is, they acquire a *noble gas outer electron configuration*. For all noble gases except helium, which has only 2 electrons, the noble gas outer electron configuration consists of eight outer electrons. The tendency of elements to acquire an 8-electron outer electron configuration is very useful in predicting the nature of chemical bonding and the formulas of compounds that result and is called the *octet rule*.

Completion of the abbreviated periodic table

The abbreviated version of the periodic table can be finished with elements 11 through 20. The atomic numbers, names, and symbols of these elements are 11, sodium, Na; 12, magnesium, Mg; 13, aluminum, Al; 14, silicon, Si; 15, phosphorus, P; 16, sulfur, S; 17, chlorine, Cl; 18, argon. Ar; 19, Potassium, K; and 20, calcium Ca. This table shows the Lewis symbols of the elements to emphasize their orderly variation across periods and similarity in groups of the periodic table. Note that, with the exception of He at the top of the last group, the configuration of dots for the Lewis symbols in each group (vertical column) are identical.

Chemical Bonds and Compounds

Most atoms are joined by *chemical bonds* to other atoms. Therefore, elemental hydrogen exists as *molecules*, each consisting of 2 H atoms and denoted by the *chemical formula*, H_2. The H atoms in the H_2 molecule are held together by a *covalent bond* made up of 2 electrons, each contributed by one of the H atoms, and shared between the atoms.

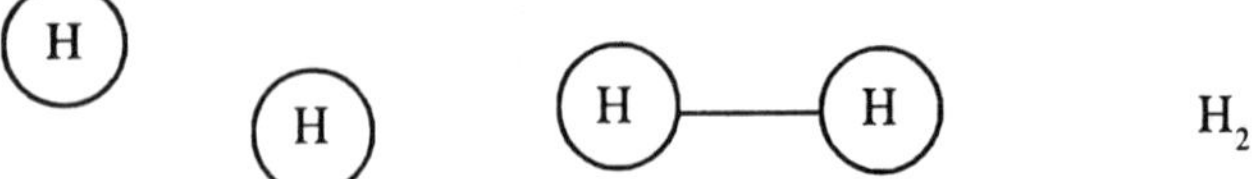

The H atoms in elemental hydrogen are held together by chemical bonds in molecules that have the chemical formula H_2.

The bonding of H atoms to form H_2 molecules may also be represented by Lewis symbols (for atoms) and Lewis formula (for molecules) as shown below. The two dots, :, in the H_2 molecule represent two shared electrons in the covalent bond holding the H molecule together.

$$H\cdot \rightleftarrows \cdot H \longrightarrow H:H$$

Fig. 4.3. Formation of a molecule of H_2 showing covalent bonds and Lewis formula of the hydrogen molecule.

Chemical Compounds

Most substances consist of two or more elements joined by chemical bonds. Hydrogen atoms combine with oxygen atoms to form molecules in which 2 H atoms are bonded to 1 O atom in a substance with a chemical formula of H_2O (water). A *chemical compound* is a substance composed of atoms of two or more different elements bonded together. In the *chemical formula* for water the subscript 2 indicates that there are 2 H atoms per O atom. (The absence of a subscript after the O denotes the presence of just 1 O atom in the molecule.) Each of the chemical bonds holding a hydrogen atom to the oxygen atom in the water molecule is formed by two electrons shared between the hydrogen and oxygen atoms.

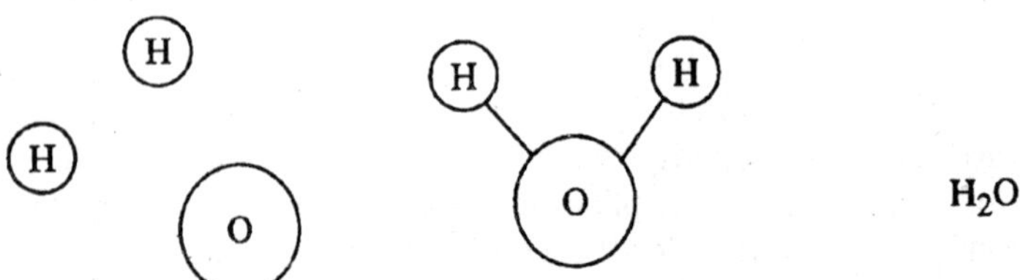

Hydrogen atoms and oxygen atoms bond together to form molecules in which 2H atoms are attached to 1 O atom. The chemical formula of the product compound, water is H_2O.

Fig. 4.4. A molecule of water, H_2O, formed from 2 H atoms and 1 O atom held together by chemical bonds.

Covalent bonds

Figure 4.5 shows covalent bonding in the formation of hydrogen compounds of C and N. In the case of carbon, 4 H atoms are combined with one C atom having 4 valence electrons, sharing electrons such

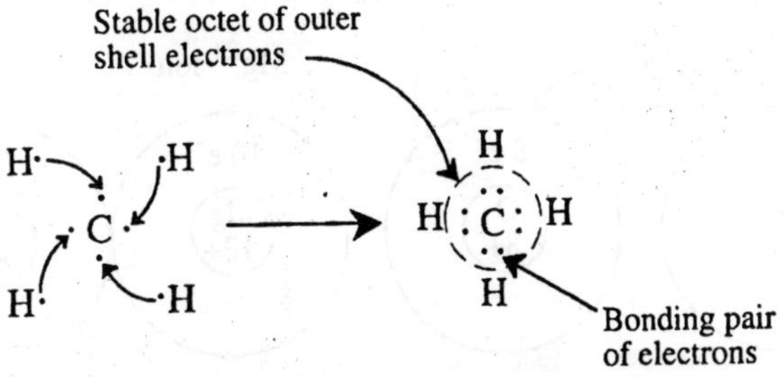

Stable octet of outer shell electrons

Bonding pair of electrons

Each of 4 H atoms shares a pair of electrons with a C atom to form a molecule of methane CH_4.

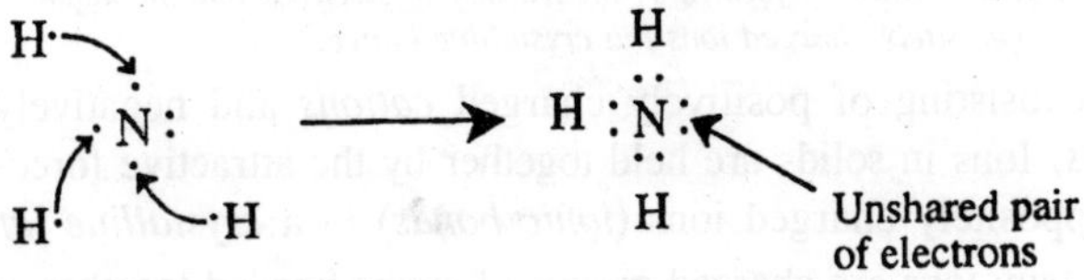

Unshared pair of electrons

Each of 3 H atoms shares a pair of electrons with an atom of N to form a molecule of ammonia, NH_3.

Fig. 4.5. Formation of stable outer electron shells by covalent bonding in compounds.

that in the CH_4 product each H atom has 2 electrons and the C atom has 8 outer shell electrons (a stable octet), all in 4 shared pairs. To form NH_3, only 3 H atoms are required to share their electrons with an atom of N having 5 outer shell electrons, leading to a compound in which the N atom has 8 outer shell electrons, 6 of which are shared with H. The *shared pairs* of electrons, in the C-H and N-H bonds shown comprise *covalent bonds.*

Lewis symbols and formulas can be used to show the atoms bound together in compounds and the types of bonds in each. The outer shell *valence electrons* are represented as dots, or each pair of valence electrons in a chemical bond is shown as a dash. The electrons in the Lewis formulas of the molecules above are shown in pairs reflecting the fact that electrons tend to be paired in groups of two in both atoms and molecules. They are said to occupy *orbitals*, each of which can contain a maximum of two electrons. The electrons in orbitals have spins in opposite directions. This is an important concept in explaining the behavior of electrons in atoms and molecules.

Ionic bonds

As shown for the formation of ionic magnesium oxide, the transfer of electrons from one atom to another produces charged species called

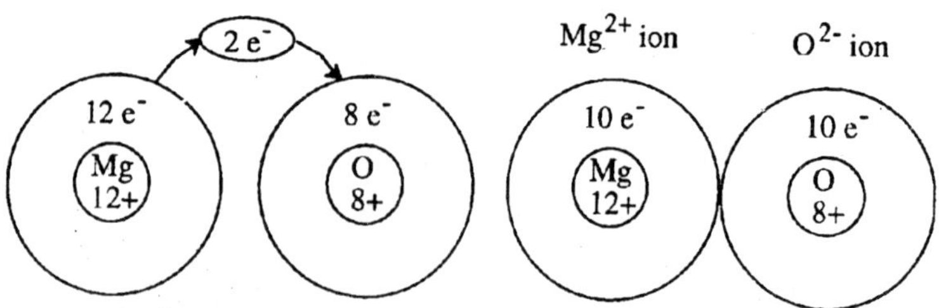

The transfer of two electrons from an atom of Mg to an O atom

yields an ion of Mg^{2+} and one of O^{2+} in the compound MgO.

Fig. 4.6. Ionic bonds are formed by the transfer of electrons and the mutual attraction of oppositely charged ions ina crystalline lattice.

ions consisting of positively charged *cations* and negatively charged *anions*. Ions in solids are held together by the attractive forces between the oppositely charged ions (*ionic bonds*) in a *crystalline lattice*.

Many ions are charged groups of atoms bonded together covalently. A common example of such an ions is the ammonium ion, NH_4^+, composed of 4 hydrogen atoms covalently bonded to a single nitrogen (N) atom and having a net electrical charge of +1 for the whole cation.

Chemical Formulas of Compounds

Chemical formulas, such as H_2 for hydrogen and H_2O for water, contain a lot of information. This is illustrated by the moderately complicated example of calcium phosphate, $Ca_3(PO_4)_2$, a compound that contains both covalent and ionic bonds. Each formula unit ("molecule") of this compound is composed of 3 Ca^{2+} ions, each with a +2 charge, and 2 phosphate ions, PO_4^{3-}, each of which has a -3 charge. The four oxygen atoms and the phosphorus atom in the phosphate ion are held together with covalent bonds. The calcium and phosphate ions are bonded together ionically in a lattice composed of these two kinds of ions.

Quantity of Matter: The Mole

The *molecular mass* (formerly called molecular weight) of a compound is calculated by multiplying the atomic mass of each element by the relative number of atoms of the element, then adding all the values obtained for each element in the compound. The molecular mass of NH_3 is $14.0 + 3 \times 1.0 = 17.0$, where 14.0 is the atomic mass of the single N atom and 1.0 is the atomic mass of each of the 3H atoms. The molecular mass of ethylene, C_2H_4, is $2 \times 12.0 + 4 \times 1.0 = 28.0$ (the atomic mass of C is 12).

Mole

The mole, used to express quantity of substance, is defined in terms of specific entities, such as atoms of Ar, molecules of H_2O, or Na^+ and Cl^- ions, each pair of which composes a "molecule" of NaCl. A *mole* is defined as the quantity of substance that contains the same number of specified entities as there are atoms of C in exactly 0.012 kg (12 g) of carbon-12. To specify the mass of a substance equivalent to its number of moles, simply state the atomic mass (of an element) or the molecular mass (of a compound) and affix "grams" to it as shown by the examples bellow:

1. *A mole of argon, which always exist as individual Ar atoms*. The atomic mass of Ar is 40.0. Therefore, exactly one mole of Ar is 40.0 grams of argon.
2. *A mole of molecular elemental hydrogen, H_2*. The atomic mass of H is 1.0, the molecular mass of H_2 is, therefore, 2.0 and a mole of H_2 contains 2.0 g of H_2.
3. *A mole of methane, CH_4*. The atomic mass of H is 1.0 and that of C is 12.0, so the molecular mass of CH_4 is 16.0. Therefore, a mole of methane has a mass of 16.0 g.

Avogadro's number (6.02×10^{23}) is the number of specified entities in a mole of substance. The "specified entities" may consist of atoms or molecules or they may be group of ions that make up the smallest possible unit of an ionic compound, such as two Na^+ ions and one S^{2-} ion in Na_2S.

Chemical Reactions and Equations

Chemical reactions occur when substances are changed to other substances. An example is the chemical reaction of hydrogen and oxygen written as the *chemical equation*,

$$2H_2 + O_2 \rightarrow 2H_2O \quad ...(1)$$

in which the arrow is read as "yields" and separates the hydrogen and oxygen *reactants* from the water *product*. All correctly written chemical equations are *balanced*, having the same number of each kind of atom on both sides of the equation. The equation above is balanced because of the following: On the left there are 2 H_2 *molecules* each containing 2 H atoms for a total of 4H atoms, and there is 1 O_2 *molecule* containing 2 O *atoms* for a total of 2 O atoms. On the right, there are 2 H_2O molecules each containing 2 H atoms and 1 O atom for a total of 4 H atoms and 2 O atoms.

Some substances enable chemical reactions to occur, but are not themselves consumed in the reactions. These materials are called *catalysts*. Catalysts are very important in chemical manufacturing. Catalysts in automotive exhaust systems enable destruction of pollutant exhaust gases. Special biological catalysts called *enzymes* enable biochemical processes to occur.

Physical Properties and State of Matter

Physical properties of matter are those that can be measured without altering the chemical composition of the matter. Three physical properties important in describing and identifying particular kinds of matter are density, color, and solubility. *Density* (d) is defined as mass per unit volume and is expressed by the formula

$$d = \frac{\text{mass}}{\text{volume}} \qquad \ldots(2)$$

The densities of liquids and solids are normally given in units of grams per cubic centimeter (g/cm^3, the same as grams per milliliter, g/mL), whereas the densities of much lighter gases are given in units of grams per liter (g/L). *Color* is of the more useful properties for identifying substances without doing any chemical or physical tests. As examples, a violet vapor is characteristic of iodine, and characteristic yellow/brown color in water may be indicative or organically bound iron. *Solubility* refers to the degree to which a substance dissolves in liquid, such as water. It is discussed along with solution later in this section.

States of Matter

Figure 4.7 illustrates the three *states of matter* in which matter may exist. *Solids* have a definite shape and volume. *Liquids* have an indefinite shape and take on the shape of the container in which they are contained. Solids and liquids are not significantly compressible,

A solid has a definite shape and volume regardless of the container into which it is placed.

A quantity of liquid has a definite volume, but takes on the shape of its container

A quantity of gas has the shape and volume of the container it occupies.

Fig. 4.7. Representatons of the three states of matter.

which means that a specific quantity of a substance has a definite volume and cannot be squeezed into a significantly smaller volume. *Gases* take on both the shape and volume of their containers. A quantity of gas may be compressed to a very small volume and will expand to occupy the volume of any container into which it is introduced.

Everyone is familiar with the three states of matter for water. These are (1) gas, such as water vapor in a humid atmosphere, steam; (2) liquid, such as water in a lake, groundwater; (3) solid, such as ice in polar ice caps, snow in snowpack. Changes in matter from one phase to another are very important in the environment, such as water vapor changing from the gas phase to liquid forming clouds or precipitation. Water is desalinated by producing water vapor from sea water, leaving the solid salt behind, and recondensing the pure water vapor as a salt-free liquid. Some organic pollutants are extracted from water for chemical analysis by transferring them from the water to another organic phase that is immiscible with water.

Gases and the Gas Laws

The atmosphere is composed of a mixture of gases, the most abundant of which are nitrogen, oxygen, argon, carbon dioxide, and water vapor. A quantity of gas takes on the shape and volume of the container in which it is held. The reason for this behavior is that gas molecules move independently and at random, bouncing off each other as they do so. Gas molecules colliding with container walls exert *pressure*. The rapid, constant motion of gas molecules explains the phenomenon of gas *diffusion* in which gases move large distances from their sources. This occurs, for example, with water vapor in the atmosphere. Evaporated gasoline can diffuse some distance from its source such that, if it contacts a flame, a fire or explosion may result.

The *general gas law* explains the relationships among the quantity of gas in numbers of moles (n), volume (V), temperature (T), and pressure (P) in the form of the *ideal gas equation*:

$$V = \frac{RnT}{P} \text{ or } PV = nRT \qquad \ldots(3)$$

For calculations involving volume in liters and pressures in atmospheres, the value of R is 0.0821 L-atm/deg-mol.

A temperature of 0°C (273.15 K) and a pressure of 1 atmosphere (atm) have been chosen as standard temperature and pressure (STP). As STP the volume of 1 mole of ideal gas is 22.4 L, a volume called the *molar volume* of a gas. The ideal gas law can be used to calculate changes in volume or other parameters resulting from changes in P,

V, n, and T. As an example, calculate the volume of 0.333 moles of gas at 300 K under a pressure of 0.950 atm:

$$V = \frac{nRT}{P} = \frac{0.333\ \text{mol} \times 0.0821\ \text{L atm / K mol} \times 300\ \text{K}}{0.950\ \text{atm}} = 8.63\ \text{L}$$

Liquids and Solutions

A given quantity of liquid occupies a fix volume, but, because of the free movement of molecules relative to each other, it takes on the shape of that portion of the container that it occupies. Molecules of liquids that are energetic enough to escape the attractive forces of the other molecules in the mass of liquid enter the gas phase, a phenomenon called *evaporation*; the opposite process is called *condensation*. Equilibrium between these two processes results in a steady-state level of vapor above a liquid called its *vapor pressure*.

Many liquids act as *solvents* to form *solutions* in which quantities of gases, solids, or other liquids are *dissolved* as *solutes*. The maximum degree to which a solute dissolves in liquid is its *solubility*. A solution that is at equilibrium with excess solute so that it contains the maximum amount of solute that it can dissolve is called a *saturated solution*. One that can still dissolve more solute is called an *unsaturated solution*.

For the chemist, the most useful way to express the concentration of a solution is in terms of the *molar concentration*, M, the number of moles of solute dissolved per liter of solution:

$$M = \frac{\text{moles of solute}}{\text{number of liters of solution}} \qquad ...(5)$$

Examples. Exactly 34.0 of ammonia, NH_3, were dissolved in water and the solution was made up to a volume of exactly 0.500 L. What was the molar concentration, M, of ammonia in the resulting solution?

Answer. The molar mass of NH_3 is 17.0 g/mole. Therefore

$$\text{Number of moles of } NH_3 = \frac{34.0\ \text{g}}{17.0\ \text{g / mole}} = 2.00\ \text{mol} \qquad ...(6)$$

$$M = \frac{2.00\ \text{mol}}{0.500\ \text{L}} = 4.00\ \text{mol / L} \qquad ...(7)$$

Solids

The *solid state* is the most organized form of matter in that the atoms, molecules, and ions in it are in essentially fixed relative positions and are highly attracted to each other. Therefore, solids have a definite shape, maintain a constant volume are virtually non-compressible under pressure, and expand and contract only slightly with changes in temperature. Because of the strong attraction of the atoms, molecules, and ions of solids for each other, solids do not enter the vapor phase readily at all; the phenomenon by which this happens to a limited extent is called *sublimation*.

Thermal Properties

The *melting point* of a pure substance is the temperature at which the substance changes from a solid to a liquid. At the melting temperature pure solid and pure liquid composed of the substance may be present together in a state of equilibrium. Boiling occurs when a liquid is heated to a temperature such that bubbles of vapor of the substance are evolved. When the surface of a pure liquid substance is in contact with the pure vapor of the substance at 1 atm pressure, boiling occurs at a temperature called the *normal boiling point*.

As the temperature of a substance is raised, energy must be put into it to enable the molecules of the substance to move more rapidly relative to each other and to overcome the attractive forces between them. The *specific heat* of a substance is defined as the amount of heat energy required to raise the temperature of a gram of substance by 1 degree Celsius. The *heat of vaporization* required to convert liquid water to vapor is 2,260 J/g (2.26 kJ/g) for water boiling at 100°C at 1 atm pressure. This amount of heat energy is about 540 times that required to raise the temperature of a 1 gram of liquid water by 1°C. When water vapor condenses, similar enormous amounts of heat energy called *heat of condensation* are released. *Heat of fusion* is the quantity of heat taken up in converting a unit mass of solid entirely to liquid at a constant temperature. The heat of fusion of water is 330 J/g for ice melting at 0°C and 80 times the specific heat of water.

Acids, Bases, and Salts

Almost all inorganic compounds and many organic compounds can be classified as acids, bases, or salts, and they are discussed in this section. Acids, bases, and salts are very important in life processes, in the environment, and as industrial chemicals.

An *acid* is a substance that produces hydrogen ions, H^+, in water. (Actually, in water, H^+ ion is associated with water molecules in clusters such as they *hydronium ion*, H_3O^+, but for simplicity in this book, H^+ in water will be shown simply as H^+.) For example, HCl in water is entirely in the form of H^+ ions and Cl^- ions. These two ions in water form hydrochloric acid. Toxic hydrocyanic acid, HCN, also produces hydrogen ions in water:

$$HCN \rightarrow H^{+} + CN^{-} \quad ...(8)$$

When the HCN molecule comes apart, it is said to *ionize*, and the process is called *ionization*. At all but extremely low HCN concentrations, only a small percentage of the acid molecules release H^{+} ion. Therefore, HCN is said to be a *weak acid*, a term that will be defined further in this section.

Bases

A *base* is a substance that produces hydroxide ion, OH^{-}, and/or accepts H^{+}. Many bases consist of metal ions and hydroxide ions. For example, solid sodium hydroxide dissolves in water

$$NaOH(s) \rightarrow Na^{+}(aq) + OH^{-}(aq) \quad ...(9)$$

to yield a solution containing OH^{-} ions. (In this equation (*s*) denotes a solid and (*aq*) stands for a substance dissolved in water; (*g*) is used to designate a gas in equations involving gases.) When ammonia gas is bubbled into water, a limited number of the molecules of NH_3 remove hydrogen ion from water and produce ammonium ion, NH_4^{+}, and hydroxide ion as shown by the following reaction:

$$NH_3 + H_2O \rightarrow NH_4^{+} + OH^{-} \quad ...(10)$$

Salts

Whenever an acid and a base are brought together, water is always a product, leaving a negative ion form the acid and a positive ion from the base:

$$\underset{\text{hydrochloric acid}}{H^{+} + Cl^{-}} + \underset{\text{sodium hydroxide}}{Na^{+} + OH^{-}} \rightarrow \underset{\text{sodium chloride}}{NA^{+} + Cl^{-}} + \underset{\text{water}}{H_2O} \quad ...(11)$$

Sodium chloride is a *salt*, a compound composed of a positively charged *cation* other than H^{+} and negatively charged *anion* other than OH^{-}.

Dissociation of Acids and Bases in Water

The reactions discussed above have shown that acids and bases dissociate, or ionize, in water to produce ions. There is great difference in how much various acids and bases ionize. Some, like HCl or NaOH, are completely dissociated in water. Because of this hydrochloric acid is called a *strong acid*. Sodium hydroxide is a *strong base*. Partially ionized acids and base, such as hydrocyanic acid and ammonia, mentioned above are *weak acids* and *weak bases*. Some common strong acids are sulfuric acid (H_2SO_4) and nitric acid (NHO_3); weak acids include acetic acid ($HC_2H_3O_2$, of which only one of the four H's can form H^+ ion), and hypochlorous acid (HCIO). The percentage of weak acid molecules that are ionized depends upon the concentration of the acid. The lower the concentration, the higher the percentage of ionized molecules.

Hydrogen Ion Concentration and pH

Hydrogen ion concentration, commonly denoted ad $[H^+]$ and expressed in moles liter, M, is a very important characteristic of some solutions. For example, the value of $[H^+]$ in human blood must stay within relatively narrow ranges, or the person will become ill or even die. Fortunately, there are mixtures of chemicals that keep the H^+ concentration of a solution relatively constant. Reasonable quantities of acid or base added to such solutions do not cause large changes in H^+ concentration. Solutions that resist changes in $[H^+]$ are called *buffers*.

Because of the fact that water, itself, produces both H^+ and hydroxide ion.

$$H_2O \leftarrow\rightarrow H^+ + OH^- \qquad \text{...(12)}$$

there is always some H^+ and some OH^- in any solution. (The reverse arrow shows that H^+ and OH^- ions recombine to give H_2O molecules.) Of course, in an acid solution the concentration of OH^- is always very low, and in a solution of base the concentration of H^+ is very low. If the value of either $[H^+]$ or $[OH^-]$ in moles per liter (M) is known, the value of the other can be calculated from the following relationship:

$$[H^+][OH^-] = 1.00 \times 10^{-14} = K_w \text{ (at 25°C)} \qquad \text{...(13)}$$

For example, in a solution of 0.100 M HCl in which $[H^+] = 0.100$ M,

$$[OH^-] = \frac{K_W}{[H^+]} = \frac{1.00 \times 10^{-14}}{0.100} = 1.00 \times 10^{-13} \text{ M} \qquad \text{...(14)}$$

Molar concentrations of hydrogen ion, $[H^+]$, range over many orders of magnitude and are conveniently expressed by pH defined as

$$pH = -\log[H^+] \qquad \text{...(15)}$$

In absolutely pure water $[H^+] = [OH^-]$, the value of $[H^+]$ is exactly 1×10^{-7} mole/L at 25°C, the pH is 7.00 and the solution is *neutral* (neither acidic nor basic). *Acidic* solutions have pH values of less than 7, and *basic* solutions have pH values of greater than 7. When the H^+ ion concentration is 1 times 10 to a power (the superscript number, such as –2, –7, etc), the pH is simply the negative value of that power. Thus, when $[H^+]$ is 1×10^{-3}, the pH is 3; when $[H^+]$ is 1×10^{-4}, the pH is 4. For a solution with a hydrogen ion concentration between 1×10^{-4} and 1×10^{-3}, such as 3.16×10^{-4}, the pH is obviously going to be between 3 and 4. The pH is calculated very easily on an electronic calculator by entering 3.16×10^{-4} on the keyboard and pressing the "log" button, which gives –3.50, so the pH is 3.50.

Organic Chemistry

Most carbon-containing compounds are *organic chemicals* and are addressed by the subject of *organic chemistry*. Organic chemistry is a vast, diverse, discipline because of the enormous number of organic compounds that exist as a consequence of the versatile bonding capabilities of carbon. Such diversity is due to the ability of carbon atoms to bond to each other through single (2 shared electrons) bonds, double (4 shared electrons) bonds, and triple (6 shared electrons) bonds, in a limitless variety of straight chains, branched chains, and rings. All organic compounds, of course, contain carbon. Virtually all also contain hydrogen and have at least one C–H bond.

Among organic chemicals are included the majority of important industrial compounds, synthetic polymers, agricultural chemicals, biological materials, and most substances that are of concern because of their toxicities and other hazards. Pollution of the water, air, and soil environments by organic chemicals is an area of significant concern.

Molecular Geometry in Organic Chemistry

The three-dimensional shape of a molecule, called its *molecular geometry*, is particularly important in organic chemistry. This is because its molecular geometry determines in part the properties of anorganic molecule, particularly its interactions with biological systems. Shapes of molecules are represented in drawings by lines of normal, uniform thickness for bonds in the plane of the paper, and with broken lines for bonds extending away from, and heavy lines for bonds extending toward the viewer. These conventions are shown by the example of dichloromethane, CH_2Cl_2.

Hydrocarbons

The simplest and most easily understood organic compounds are *hydrocarbons*, which contain only hydrogen and carbon. The major types of hydrocarbons are alkanes, alkenes, alkynes, and aryl compounds. In the structures shown, C=C represents a *double bond* in which four electrons are shared, C=C is a *triple bond* in which six electrons are shared.

Alkanes

Alkanes, also called *paraffins* or *aliphatic hydrocarbons*, are hydrocarbons in which the C atoms are joined by single covalent bonds

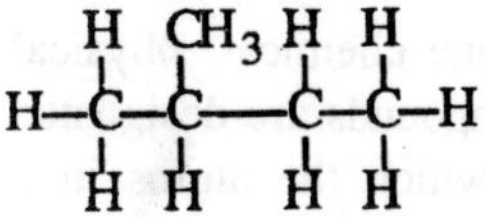

H–C≡C–H

2-Methylbutane (alkane)

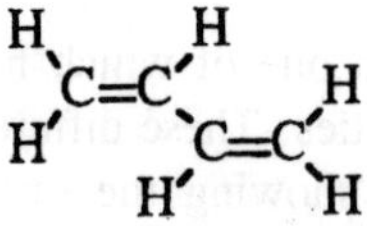

1,3-Butadiene (alkene)

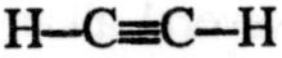

Acetylene (alkyne)

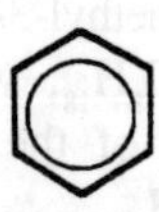

Benzene (aryl compound)

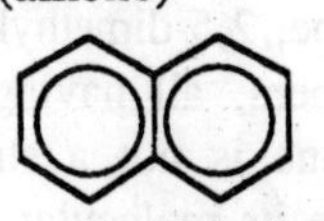

Naphthalene (aryl compound)

Fig. 4.8. Examples of major types of hydrocarbons.

(sigma bonds) consisting of two shared electrons. The three major kinds of alkanes are, respectively, *straight-chain alkanes*, *branched-chain alkanes*, and *cycloalkanes*.

Formulas of alkanes

Formulas of organic compounds present information at several different levels of sophistication. *Molecular formulas*, such as that of octane (C_8H_{18}), give the number of each kind of atom in a molecule of a compound. However, the molecular formula of C_8H_{18} may apply

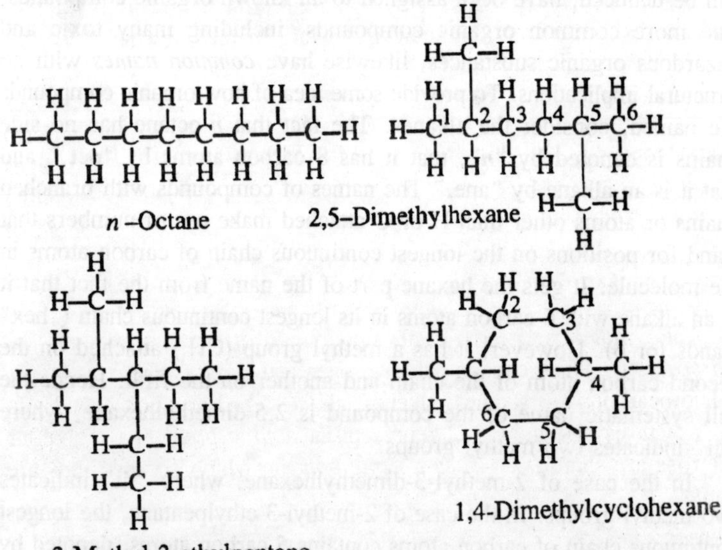

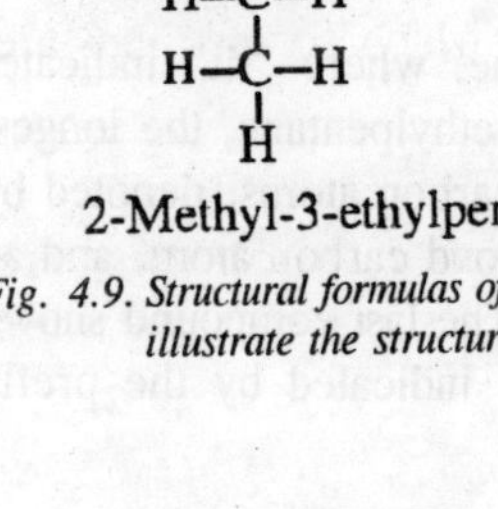

Fig. 4.9. Structural formulas of four hydrocarbons, each containing 8 carbon atoms, that illustrate the structural diversity possible with organic compounds.

to several alkanes, each one of which has unique chemical, physical and toxicological properties. These different compounds are designated by *structural formulas* showing the order in which the atoms in a molecule are arranged. Compounds that have the same molecular, but different structural, formulas are called *structural isomers*. Of the compounds, *n*-octane, 2,5-dimethylhexane, and 2-methyl-3-ethylpentane are structural isomers, all having the formula C_8H_{18}, whereas 1,4-demethylcyclohexane is not a structural isomer of the other three compounds because its molecular formulas is C_8H_{16}.

Alkanes and alkyl groups

Most organic compounds can be derived from alkanes, and many important parts of organic molecules contain one or more alkane groups minus a hydrogen atom bonded as substituents onto the basic organic molecule. As a consequence, the names of many organic compounds are based upon alkanes. Two important substituent groups derived from alkanes are the methyl group, $-CH_3$, (derived from methane, CH_4) and ethyl group, $-C_2H_5$, (derived from ethane, C_2H_6).

Names of alkanes and organic nomenclature

Systematic names, from which the structures or organic molecules can be deduced, have been assigned to all known organic compounds. The more common organic compounds, including many toxic and hazardous organic substances, likewise have *common names* with no structural implications. To provide some idea of how organic compounds are named, consider the alkanes. The fact that *n*-octane has no side chains is denoted by "*n*", that it has 8 carbon atoms by "oct," and that it is an alkane by "ane." The names of compounds with branched chains or atoms other than H or C attached make use of numbers that stand for positions on the longest continuous chain of carbon atoms in the molecule. It gets the hexane part of the name from the fact that it is an alkane with 6 carbon atoms in its longest continuous chain ("hex" stands for 6). However, it has a methyl group (CH_3) attached on the second carbon atom of the chain and another on the fifth. Hence the full systematic name of the compound is 2,5-dimethylhexane, where "di" indicates two methyl groups.

In the case of 2-methyl-3-dimethylhexane, where "di" indicates two methyl groups. In the case of 2-methyl-3-ethylpentane, the longest continuous chain of carbon atoms contains 5 carbon atoms, denoted by pentane; a methyl group is attached to the second carbon atom, and an ethyl group, C_2H_5, on the third carbon atom. The last compound shown in the figure has 6 carbon atoms in a ring, indicated by the prefix "cyclo," so it is a cyclohexane compound. Furthermore, the carbon in the ring to which one of the methyl groups is attached is designated by "a" and another methyl group is attached to the fourth carbon atom around the ring. Therefore, the full name of the compound is 1,4-dimethylcyclohexane.

The basic rules used in naming simple alkanes are the following: (1) The name of the compound is based upon the longest continuous chain of carbon atoms, (2) the carbon atoms in the longest continuous chain are numbered sequentially from one end, (3) all groups attached to the longest continuous chain are designated by the number of the carbon atom to which they are attached and by the name of the substituent group, (4) a prefix is used to denote multiple substitutions by the same kind of group, and (5) the complete name is assigned such that it denotes the longest continuous chain of carbon atoms and the name and location on this chain of each substituent group.

Reactions of alkanes

Alkanes are relatively unreactive. At elevated temperatures they readily burn with molecular oxygen in air as shown by the following reaction of propane:

$$C_3H_8 + 5O_2 \rightarrow 3CO_2 + 4H_2O + \text{heat} \quad \text{...(16)}$$

Common alkanes are highly flammable and the more volatile lower molecular mass alkanes from explosive mixtures with air.

In addition to combustion, alkanes undergo *substitution reactions* in which one or more H atoms on an alkane are replaced by atoms of another element. The most common such reaction is the replacement of H by chlorine, to yield *organochlorine* compounds. For example, methane reacts with chlorine to give chloromethane, as shown below

$$Cl_2 + CH_4 \rightarrow CH_3Cl + HCl \quad \text{...(17)}$$

Alkanes and Alkynes

Alkenes (olefins) are hydrocarbons that have double bonds consisting of 4 shared electrons. The simplest and most widely manufactured alkene is ethene (ethylene),

$$\begin{matrix} H & & H \\ & C{=}C & \\ H & & H \end{matrix} \quad \textbf{Ethylene (ethene)}$$

used for the production of polyethylene polymer. Another example of an important alkene is, 1,3-butadiene widely used in the manufacture of polymers, particularly synthetic rubber.

Acetylene is an *alkyne*, a class of hydrocarbons characterized by carbon-carbon triple bonds consisting of 6 shared electrons. Highly flammable, dangerously explosive acetylene is used in large quantities as a chemical raw material and fuel for oxyacetylene torches.

Addition reactions

The double and triple bonds in alkenes and alkynes have "extra" electrons capable of forming additional bonds. Therefore, the carbon atoms attached to these bonds can add atoms without losing any atoms already bonded to them, and the multiple bonds are said to be *unsaturated*. Therefore, alkenes and alkynes both under *addition reaction* in which pairs of atoms are added across unsaturated bonds as shown in the hydrogenation reaction of ethylene with hydrogen to give ethane:

$$H_2C{=}CH_2 + H{-}H \rightarrow H_3C{-}CH_3$$

Addition reactions, which are not possible with alkanes, add to the chemical and metabolic versatility of compounds containing unsaturated bonds and constitute a factor contributing to their generally higher toxicities. Addition reactions make unsaturated compounds much more chemically reactive, more hazardous to handle in industrial processes, and more active in atmospheric chemical processes, such as smog formation.

Alkenes and *Cis-trans* Isomerism

As shown by the two simple compounds, the two carbon atoms connected by a double bond in alkenes cannot rotate relative to each other. For this reason, another kind of isomerism, called *cis-trans* isomerism is possible for alkenes.

$H_3C(H)C{=}C(H)CH_3$ (cis)	$H_3C(H)C{=}C(CH_3)H$ (trans)
Cis-2-butene, both CH_3 groups on the same side of the double bond	*Trans*-2-butene, both CH_3 groups on opposite sides of the double bond

Fig. 4.10. Cis and trans isomers of the alkene, 2-butene, C_4H_8.

Aryl Hydrocarbons

Benzene is the simplest of a large class of *aryl* (*aromatic*) hydrocarbons. Many important aryl compounds have substituent groups containing atoms of elements other than hydrogen and carbon and are called *aryl compounds* or *aromatic compounds*. Most aryl compounds

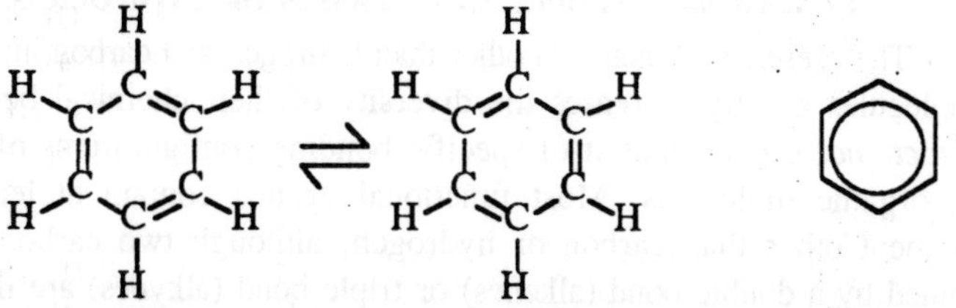

Fig. 4.11. Representation of the aryl benzene molecule with two resonance structures (left) and more accurately, as a hexagon with a circle in it (right).

contain 6-carbon-atom benzene rings as shown for benzene, C_6H_6. The atoms in aryl compounds are held together in part by particularly stable bonds that contain delocalized clouds of so-called π (pi, pronounced "pie") electrons. In an oversimplified sense the structure of benzene can be visualized as resonating between the two equivalent structures by the shifting of electrons in chemical bonds. This structure can be shown more simply and accurately by a hexagon with a circle in it.

Many toxic substances, environmental pollutants, and hazardous waste compounds, such as benzene, toluene, naphthalene, and chlorinated phenols, are aryl compounds. Some aryl compounds, such as naphthalene and the polycyclic aromatic compound, benzo(a)pyrene, contain fused rings.

Benzo(a)pyrene is the most studied of the polycyclic aryl hydrocarbons (PAHs), which are characterized by condensed ring systems ("chicken wire" structures). These compounds are formed by the incomplete combustion of other hydrocarbons. Some PAH compounds, including benzo(a)pyrene, are of toxicological concern because they are precursors to cancer-causing metabolites.

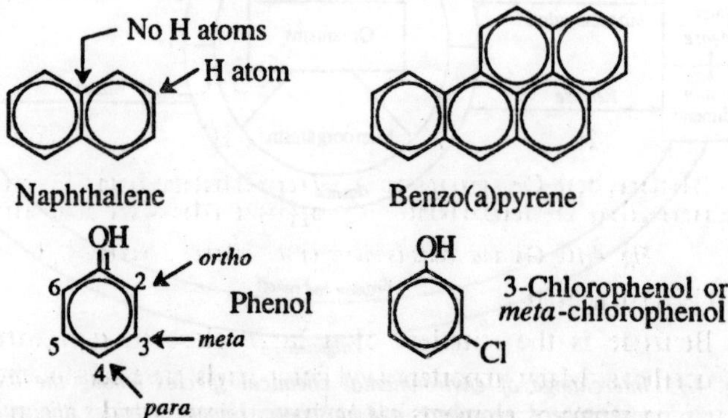

Fig. 4.12. Aryl compounds containing fused rings (top) and showing the numbering of carbon atoms for purposes of nomenclature.

Functional Groups and Classes of Compounds

The presence of elements other than hydrogen and carbon in organic molecules greatly increases the diversity of their chemical behavior. *Functional groups* consist of specific bonding configurations of atoms in organic molecules. Most functional groups contain at least one element other than carbon or hydrogen, although two carbon atoms joined by a double bond (alkenes) or triple bond (alkynes) are likewise considered to be functional groups.

Nonhydrocarbon organic compounds may be classified according to the elements other than carbon and hydrogen in them. Important types of such compounds are organooxygen compounds, organonitrogen compounds, organohalides, organosulfur compounds, and organophosphorus compounds.

Environmental Chemistry

As shown graphically by the interchange of chemical species among various environmental spheres, environmental chemistry is the study of the sources, reactions, transport, effects, and fates of chemical species in the water, air, terrestrial and living environments and the effects of human activities thereon.

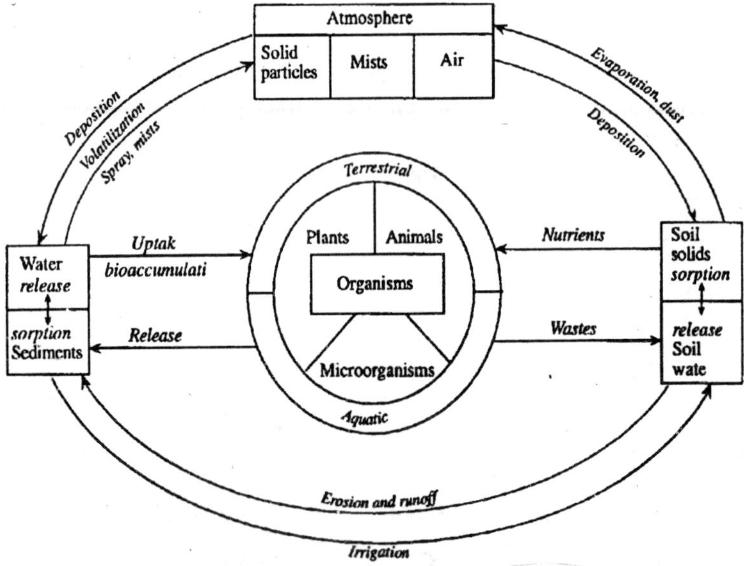

Fig. 4.13. Interchange of environmental chemical species among the atmosphere, hydrosphere, geosphere, and biosphere. Human activities (the anthrosphere) have a strong influence on the various processes shown.

AQUATIC CHEMISTRY

Water has a number of unique properties that are essential to life, due largely to its molecular structure and bonding properties. Among the special characteristics of water are the fact that it is an excellent solvent, it has a temperature/density relationship that results in bodies of water becoming stratified in layers, it is transparent, and it has extraordinary capacity to absorb, retain, and release heat per unit mass of ice, liquid water, or water vapor.

Figure 4.14 summarizes the more important aspects of *aquatic chemistry* as it applies to environmental chemistry. As shown in this figure, a number of chemical phenomena occur in water. Many aquatic chemical processes are influenced by the action of algae and bacteria in water. For example, it is shown that algal photosynthesis fixes inorganic carbon from HCO_3^- ion in the form of biomass (represented as $\{CH_2O\}$), in a process that also produces carbonate ion, CO_3^{2-}. Carbonate undergoes an acid-base reaction to produce OH^- ion and raise the pH, or it reacts with Ca^{2+} ion to precipitate solid $CaCO_3$. Most of the many oxidation-reduction reactions that occur in water are mediated (catalyzed) by bacteria. For example, bacteria convert inorganic nitrogen largely to ammonium ion, NH_4^+, in the oxygen-deficient (anaerobic) lower layers of a body of water. Near the surface, which is aerobic because O_2 is available, bacteria convert inorganic nitrogen to nitrate ion, NO_3^-. Metals in water may be bound to organic chelating agents, such as pollutant nitrilotriacetic acid (NTA) or naturally occurring fulvic acids produced by decay of plant matter. Gases are exchanged with the atmosphere, and various solutes are exchanged between water and sediments in bodies of water.

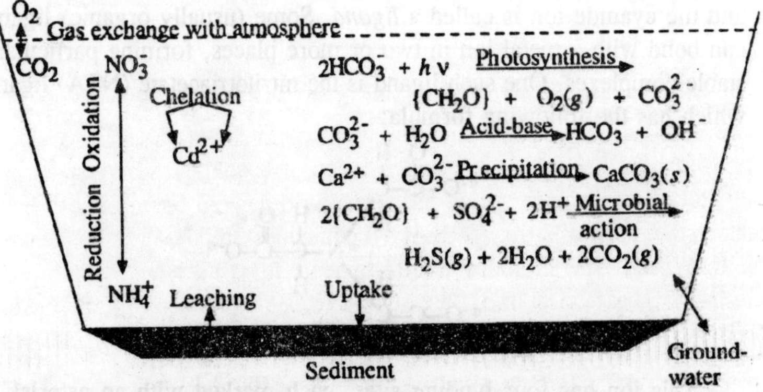

Fig. 4.14. Major aquatic chemical processes.

Oxidation-reduction

Oxidation-reduction (redox) reactions in water involve the transfer of electrons between chemical species. In natural water, wastewater, and soil, most significant oxidation-reduction reactions are carried out by bacteria.

The relative oxidation-reduction tendencies of a chemical system depend upon the *activity of the electron* e^-. When the electron activity is relatively high, chemical species, including water, tend to accept electrons,

$$2H_2O + 2e^- \leftarrow\rightarrow H_2(g) + 2OH^- \qquad ...(18)$$

and are said to be *reduced*. When the electron activity is relatively low, the medium is *oxidizing*, and chemical species such as H_2O may be *oxidized* by the loss of electrons:

$$2H_2O \leftarrow\rightarrow O_2(g) + 4H^+ + 4e^- \qquad ...(19)$$

The relative tendency toward oxidation or reduction may be expressed by the electrode potential E, which is more positive in an oxidizing medium and more negative in a reducing medium.

Complexation and Chelation

Metal ions in water are always bonded to water molecules in the form of hydrated ions represented by the general formula. $M(H_2O)_x^{n+}$, from which the H_2O is often omitted for simplicity. Other species may be present that bond to the metal ion more strongly than does water. For example, cadmium ion dissolved in water, Cd^{2+}, reacts with cyanide ion, CN^-, as follows:

$$Cd^{2+} + CN^- \rightarrow CdCN^+ \qquad ...(20)$$

The product of the reaction is called a *complex* (or complex ion) and the cyanide ion is called a *ligand*. Some (usually organic) ligands can bond with a metal ion in two or more places, forming particularly stable complexes. One such ligand is the nitrilotriacetate (NTA) ligand, which has the following formula:

```
     O H
     ‖ |
 *-O-C-C
       | \         H O
       H  \        | ‖
           *N------C-C-O-*
          /        |
       H /         H
       |/
 *-O-C-C
     ‖ |
     O H
```

This ion has four binding sites, each marked with an asterisk in the preceding illustration, which may simultaneously bond to a meatal ion, forming a structure with three rings. Such a species is known as a *chelate*, and NTA is a *chelating* agent. The most important class of complexing agents that occur naturally are the *humic substances*, degradation-resistant materials formed during the decomposition of vegetation and found in water, sediments, and soil.

In addition to metal complexes and chelates, another major type of environmentally important metal species consists of *organometallic compounds*. These differ from complexes and chelates in that the organic portion is bonded to the metal by a carbon-metal bond and the organic ligand is frequently not capable of existing as a stable separate species.

Complexation, chelation, and organometallic compounds formation have strong effects upon metals in the environment. For example, complexation with negatively charged ligands may convert a soluble metal species from a cation, which is readily bound and immobilized by ion exchange processes in soil, to an anion, such as $Ni(CN)_4^{2-}$, that is not strongly held by soil. On the other hand, some chelating agents are used for the treatment of heavy metal poisoning, and insoluble chelating agents, such as chelating resins, can be used to remove metals from waste streams.

Water Interactions with Other Phases

Most of the important chemical phenomena associated with water do not occur in solution, but rather through interaction of solutes in water with other phases. Such interactions may involve exchange of solute species between water and sediments, gas exchange between water and the atmosphere, and effects of organic surface films. Substances dissolve in water from other phases, and gases are evolved and solids precipitated as the result of chemical and biochemical phenomena in water. Sediments are repositories of a wide variety of chemicals species and the site of many chemical and biochemical processes. Sediments are sinks for many hazardous organic compounds and heavy metal salts that have gotten into water.

Colloids, which consist of very small particles ranging from 0.001 micrometer (μm) to 1 μm in diameter, have a strong influence on aquatic chemistry. Colloids have very high surface-to-volume ratios, so that they can be very active physically, chemically, and biologically. Colloids may be very difficult to remove from water during water treatment.

Water Pollutants

Natural waters are afflicted with a wide variety of inorganic, organic, and biological pollutants. In some cases, such as that of highly toxic cadmium, a pollutant is directly toxic at a relatively low level. In other cases the pollutant itself is not toxic, but its presence results in conditions detrimental to water quality. For example, biodegradable organic matter in water is often not toxic, but the consumption of oxygen during its degradation prevents the water from supporting fish life. Some contaminants, such as NaCl, are normal constituents of water at low levels, but harmful pollutants at higher levels.

Water Treatment

The treatment of water can be considered under the two major categories of (1) treatment before use and (2) treatment of contaminated water after it has passed through a municipal water system or industrial process. In both cases, consideration must be given to potential contamination by pollutants.

Several operations may be employed to treat water prior to use. Aeration is used to drive off odorous gases, such as H_2S, and to oxidize soluble Fe^{2+} and Mn^{2+} ions to insoluble forms. Lime is added to remove dissolved calcium (water hardness). $Al_2(SO_4)_3$ forms a sticky precipitate of $Al(OH)_3$, which causes very fine particles to settle. Various filtration and settling processes are employed to treat water, Chlorine, Cl_2, is added to kill bacteria.

Municipal wastewater may be subjected to primary, secondary, or advanced water treatment. *Primary* water treatment consists of settling and skimming operations that remove grit, grease, and physical objects from water. *Secondary* water treatment is designed to take out biochemical oxygen demand, BOD. This is normally accomplished by introducing air and microorganisms such that waste biomass in the water, $\{CH_2O\}$, is removed by aerobic respiration:

$$\{CH_2O\} + O_2 \rightarrow CO_2 + H_2O \qquad \ldots(21)$$

Geosphere and Soil

The *geosphere*, or solid Earth, is that part of the Earth upon which humans live and from which they extract most of their food, minerals, and fuels. Once thought to have an almost unlimited buffering capacity against the perturbations of humankind, the geosphere is now known to be rather fragile and subject to harm by human activities, such as mining, acid rain, erosion from poor cultivation practices, and disposed of hazardous wastes.

It may be readily seen that the preservation of the geosphere in a form suitable for human habitation is one of the greatest challenges facing humankind.

Soil

Soil consists of a large variety of material composing the uppermost layer of the earth's crust upon which plants grow. In addition to solids, soil contains air and water. Typically, soil solids consist of about 95% mineral matter and 5% organic material, although the proportions vary widely. Soils are formed by the weathering (physical and chemical disintegration) of parent rocks as the result of interactive geological, hydrological, and biological processes. Soils are porous and are vertically stratified into *horizons* through the action of water, organisms, and weathering processes. Soils are open systems that undergo continual exchange of mater and energy with the atmosphere, hydrosphere, and biosphere. The most active and important part of soil is *topsoil*, the layer in which plants are rooted and in which most biological activity occurs.

Atmosphere and Atmospheric Chemistry

The atmosphere consists of the thin layer of mixed gases covering the earth's surface. Exclusive of water, atmospheric air is 78.1% (by volume) nitrogen, 21.0% oxygen 0.9% argon, and 0.03% carbon dioxide. Normally, air contains 1-3% water vapor by volume. In addition, air contains a large variety of trace level gases at levels below 0.002%, including neon, helium, methane, krypton, nitrous oxide, hydrogen, xenon, sulfur dioxide, ozone, nitrogen dioxide, ammonia, and carbon monoxide.

The atmosphere is divided into several layers on the basis of temperature. Of these, the most significant are the troposphere extending in altitude from the earth's surface to approximately 11 kilometers (km) and the stratosphere from bout 11 km to approximately 50 km. The temperature of the troposphere ranges from an average of 15°C at sea level to an average of -56°C at its upper boundary. The average temperature of the stratosphere increases from -56°C at its boundary with the troposphere to -2°C at its upper boundary. The reason for this increase is absorption of solar ultraviolet energy by ozone (O_3) in the stratosphere.

The most significant feature of atmospheric chemistry is the occurrence of *photochemical reactions* resulting form the absorption by molecules of light photons, designated $h\nu$. (The energy, E, of a photon of visible or ultraviolet light is given by the equation, $E = h\nu$, where h is Planck's constant and ν is the frequency of light, which is inversely proportional to its wavelength. Ultraviolet radiation has a higher frequency than visible light and is, thereon, more energetic and more likely to break chemical bonds in molecules that absorb it.) One of the most significant photochemical reactions is the one responsible for the presence of ozone in the troposphere, which is initiated when O_2 absorbs highly energetic ultraviolet radiation in the wavelength ranges of 135-176 nanometers (nm) and 240-260 nm in the stratosphere:

$$O_2 + h\nu \rightarrow O + O \qquad \text{...(22)}$$

The oxygen atoms produced by the photochemical dissociation of O_2 react with oxygen molecules to produce ozone, O_3.

$$O + O_2 + M \rightarrow O_3 + M \qquad \text{...(23)}$$

where M is a third body, such as a molecule of N_2, which absorbs excess energy from the reaction. The ozone that is formed is very effective in absorbing ultraviolet radiation in the 220-230 nm wavelength range, which causes the temperature increase observed in the atmosphere. The ozone serves as a very valuable filter to remove ultraviolet radiation from the sun's rays. If this radiation reached the earth's surface, it would cause skin cancer and other damage to living organisms.

Gaseous Oxides in the Atmosphere

Oxides of carbon, sulfur, and nitrogen are important constituents of the atmosphere and the pollutants at higher levels. Of these, carbon dioxide, CO_2, is the most abundant. It is a natural atmospheric constituent, and it is required for plant growth. However, the level of carbon dioxide in the atmosphere, now at about 350 parts per million (ppm) by volume, is increasing by about 1 ppm per year. This increase in atmospheric CO_2 may well cause general atmospheric warming—the "Greenhouse effect," with potentially very serious consequences for the global atmosphere and for life on earth. Though not a global threat, carbon monoxide, CO, can be a serious health threat because it prevents blood from transporting oxygen to body tissues.

The two most serious nitrogen oxide air pollutants are nitric oxide, NO, and nitrogen dioxide, NO_2, collectively denoted ad "NO_x." These tend to enter the atmosphere as NO, and photochemical processes in the atmosphere tend to convert NO to NO_2. Further reactions can result in the formation of corrosive nitrate salts or nitric acid. HNO_3. Nitrogen dioxide is particularly significant in atmospheric chemistry because of its photochemical dissociation by light with a wavelength less than 430 nm to produce highly reactive O atoms. This is the first step in the formation of photochemical smog. Sulfur dioxide, SO_2, is a reaction product of the combustion of sulfur-containing fuels, such as high-sulfur coal. Part of this sulfur dioxide is converted in the atmosphere to sulfuric acid, H_2SO_4, normally the predominant contributor to acid precipitation.

Hydrocarbons and Photochemical Smog

The most abundant hydrocarbon in the atmosphere is methane, chemical formula CH_4. This gas is released from underground source as natural gas and produced by the fermentation of organic matter. Methane is one of the least reactive atmospheric hydrocarbons and is produced by diffuse sources, so that its participation in the formation of pollutant photochemical reaction products is minimal. The most significant atmospheric pollutant hydrocarbons are the reactive ones produced as automobile exhaust emissions. In the presence of NO, under conditions of temperature inversion, low humidity, and sunlight, these hydrocarbons produce undesirable *photochemical smog* manifested by the presence of visibility-obscuring particulate matter, oxidants such as ozone, and noxious organic species such as aldehydes.

Particulate Matter

Particles ranging from aggregates of a few molecules to pieces of dust readily visible to the naked eye are commonly found in the atmosphere. Some of these particles, such as sea slat formed by the evaporation of water from droplets of sea spray, are natural and even beneficial atmospheric constituents. Very small particles called *condensation nuclei* serve as bodies for atmospheric water vapor to condense upon and are essential for the formation of precipitation.

Colloidal-sized particles in the atmosphere are called *aerosols*. Those formed by grinding up bulk matter are known as *dispersion aerosols*, whereas particles formed from chemical reactions of gases are *condensation aerosols*; the latter tend to be smaller. Smaller particles are in general the most harmful because they have a greater tendency to scatter light and are the most respirable (tendency to be inhaled into the lungs).

Much of the mineral particulate matter in a polluted atmosphere is in the form of oxides and other compounds produced during the combustion of high-ash fossil fuel. Smaller particles of *fly ash* enter furnace flues and are efficiently collected in a properly equipped stack system. However, some fly ash escapes through the stack and enters the atmosphere. Unfortunately, the fly ash thus released tends to consist of smaller particles that do the most damage to human health, plants, and visibility.

5

Water Pollution

Freshwater is a scarce and valuable resource—one that can easily be contaminated. Once contaminated to the extent it can be considered "polluted," freshwater quality is difficult and expensive to restore. Thus, the study of surface water pollution has focused primarily on streams and lakes, and most of the scientific tools developed by such regulatory agencies as the U.S. Environmental Protection Agency have been applied to protecting water quality in this segment of earth's surface waters.

The water stored in reservoirs and lakes, together with the water that flow perennially in streams, is subject to heavy stress, and because it is used for water supplies, agriculture, industry, and recreation, this water can easily be contaminated.

Marine Water Resources

Oceans contain most of the water of the planet. Yet even with the phenomenal volume of water in which contaminants may dispersed, marine resources can be polluted. Using various biological and physical parameters, we usually classify the ocean environment as three components: the coastal zone, the upper mixed layer, and the abyssal ocean. Several regulatory agencies and international organizations share different responsibilities for those components. The coastal zone, which is most susceptible to the day-to-day kinds of contamination found in freshwater lakes and rivers, is often the province of water quality regulatory agencies established by individual nation states. International organizations have traditionally dealt with pollution concerns of the open ocean and its seabed, which includes the other two components. In addition to these physically described components of the sea, there are legally defined zones, sometimes overlapping, that influence regulatory practices.

Coastal Zone

The *coastal zone* extends form the low-tide line to the 200-meter depth contour, tending to match the geophysical demarcation of the continental shelf. The coastal zone can be as wide as 1400 km along some costs and less than a kilometer along others. The average width of the zone worldwide is about 50 kilometers, comprising about 8% of the surface of the ocean. (The coastal zone of Alaska is larger than that of the rest of the United States). Within the coastal zone definition, the difference between estuaries and the open coast is important in considering the disposal of wastewater and the potential for pollution problems.

Almost all of the water-carries wastes of a continent enter the coastal zone through an estuary. *Estuaries* are bodies of water with a free connection to the sea whose salinity is measurably diluted with fresh water, as from a river. Because estuaries provide critical and limited habitat for marine organisms to rear and feed their young, water quality is of special concern. Species that inhabit the coastal zone, and especially the estuaries, have to be very resilient to such natural environmental stresses as wide daily variations in salinity, turbidity, temperature, and UV radiation. Owing to this natural resiliency, coastal organisms may be able to tolerate contaminants associated with industrial and municipal wastes better than residents of the continental shelf, where the natural environment is quite stable. The estuarine habitat must be maintained primarily because of its limited extent, as distinguished from the shelf habitat, which is enormous. For this reason, treated wastewater effluents are usually discharged offshore rather than into estuaries in coastal regions. A large pipeline or tunnel, called on *outfall*, is used to transport the effluent to the disposal site.

In disposing of treated wastes offshore, we also need to take the physical features of the coastal zone into account. For example, continental headlands that protrude into the sea can impede both circulation of water and exchange of nearshore water with open ocean water. Outfalls are therefore best located far offshore rather than inside the region of headland influence. Similarly, outfalls should not be located close to shore in the vicinity of estuaries or bays because tidal incursions can carry diluted wastes into the estuary, thereby eroding one of the advantages offered by offshore disposal.

Open Ocean Waters

A variety of the majority of circumstances can contribute to the contamination of the open ocean waters beyond the coastal zone, including atmospheric fallout, oil spills, and dumping of hazardous wastes and sewage sludge as practiced by some countries of the world. Floatable and soluble materials tend to stay in the *upper mixed layer* of the ocean, where they may be decomposed. This upper layer is also the most active photosynthetic zone of the ocean, where the majority of plant—and hence animal—life can thrive. The depth of this layer, which varies between 100 and 1,000 meters, changes with season and geographic location. Although mixing between the upper and deeper layers of the ocean is impeded by strong density gradients, particles formed in the upper mixed layer, or discharged there, may eventually settle so far that they can no longer be resuspended by surface-generated turbulence and thus become part of the detrital sediment load of the deep ocean waters.

Because the quality of the water in the upper mixed layer can significantly affect all life there, it is important to take precautions with waste disposal operations. When ocean disposal of certain materials is justified, we can use technologies to avoid contamination of the upper mixed layer and facilitate transit and long-term retention of the material in the deep waters of the open ocean, that is, the *abyssal ocean*. For example, containers have been proposed for disposal of such materials as xenobiotic chemicals or radioactive wastes. Pipelines can also be used to carry liquid carbon dioxide to the seabed, where it can be retained for a long time—conceivably long enough to help reduce the rate of global warming.

Sources of Surface Water pollution

Water pollution is a qualitative term that describes the situation when the level of contaminants impedes an intended water use. It takes just a small amount of contaminated to pollute a waterbody intended for a drinking water supply. But the same water might not be considered polluted if the water were to be used, for example, for agriculture. Nor is pollution restricted to chemical contaminants. Physical factors of the environment can also contribute to pollution. For example, heated water discharged form a power plant can change the temperature of an aquatic environment. It might not be a problem in a lake or a river during the winter, but it can certainly be a problem in the summertime. Moreover, heated water or water contaminating some contaminant may not be a problem at any time of the year, provided it is rapidly mixed with the surface water, and the diluted material does not accumulate overtime. There are also many kinds of contaminants that can usually be accommodated by the natural environment without resulting in pollution, but in many situations, these same contaminants (sometimes in conjunction with other contaminants) can cause pollution even in well-mixed water bodies.

Major sources of surface water contamination are construction, municipalities, agriculture, and industry, however, the water delivered to earth in the form of precipitation is not necessarily pure to begin with. Near the coast, it may contain particulate and dissolved sea salts, and farther inland, it may contain organic compounds and acids scrubbed from contaminants added to the atmosphere both by natural processes and by anthropogenic (human) activities. Gases from plant growth and decay, and gases from geological activity are examples of naturally derived atmospheric contaminants that can be returned to earth via precipitation. The acid rain problem of the New England states is a classic example of anthropogenically derived atmospheric contaminants that contribute to surface water pollution.

Sediments as Surface Water Contaminants

The ability of rivers to carry sediment over large distances has resulted in the landscape of continents. Certainly, some background level of sediment load in rivers is considered natural and desirable. Problems ensue when anthropogenic activities in a river's watershed increase, or in some cases decrease, sediment load. Running water, wind and ice are the major factors responsible for the detachment, entrainment, and transport of particulate matter. Geologic erosion is highest in areas with relatively steep gradients such as low- to intermediate-order streams in mountainous areas. Historically, natural erosion has been the largest source of sediment supplied to rivers. As human land use activities increase in watersheds around the globe, anthropogenic effects now are major contributors to both increases sediment supplied to rivers as well as the blocking and impoundment of this sediment behind dams. Both impoundment and increased erosional processes in watersheds can have profound biological, physical, and chemical impacts on rivers and streams.

Almost any kind of human activity in watersheds can result in an increase of suspended sediment in rivers. A few classic examples of anthropogenic activity known to increase sedimentation are:

1. *Logging, deforestation, wildfire*. Specific types of logging activity can increase sediment yield by two orders of magnitude for short

periods. Fire suppression and drought can combine to create catastrophic wildfires, which can have devastating impacts on receiving waters from these areas.

2. *Overgrazing by domestic animals*. Sedimentation can increase not only due to decreased vegetative interception of precipitation-enhancing erosion, but also through direct trampling of the streambed and channel.
3. *Urbanization and road construction*. Road construction commonly results in a 5–20 fold increase in suspend sediment yield. Impervious materials such as pavement, parking lots, or rooftops can increase the velocity of storm water runoff, which will increase erosion once this water comes in contact with soil.
4. *Mining operations*. Mines, particularly strip mines, can lead to extraordinarily high levels of erosion and subsequent sedimentation in rivers. An example is the coal strip mines in Kentucky.

On a global scale, rivers discharge roughly 40,000,000 m^3 into the world's oceans annually. For every cubic meter of water reaching the ocean, there is (on average) an accompanying 0.5 kilogram of sediment carried away from the continents.

Suspended sediment is also a major carrier of pollution. While rivers may be transporters of pollution, suspended sediment is the "package" these pollutants are carried in Heavy metals, organic pollutants, pathogens, and nutrients responsible for eutrophication can also be found attached to sediments in flowing water. The "quality" of overall pollutant load of suspended sediment depends on the degree of pollution in the watershed.

Transport of sediments in water is dependent on many factors, including sediment particles size and water flow rate. The quantification and predicted rates of transport of sediment are based upon the assumption that for any given flow and sediment, there is unique transport rate. Estimates of sediment transport rates are based upon measures of flow (including velocity, depth, shear velocity, viscosity, and fluid density) and both sediment size and density. There are different classification terminologies, based upon the mode of sediment transport in a stream. *Bed load* refers to the sediments moving predominantly in contact with or close to the streambed. In contrast, *suspended load* refers to sediments that move primarily suspended in fluid flow, but that may also interact with bed load. Suspended load has a continual exchange between sediment in fluid flow and on the bed as it is constantly being entrained from the bed and suspended, while heavier particles settle out form the flow to the bed. *Solute load* refers to the total amount of dissolved materials (ions) carried in suspension and can only be quantified by laboratory analytical techniques. *Total load* is the total amount of sediment in motion and is the sum of bed load plus suspended load. It is important to remember that these classifications are somewhat artificial. The sediment load carrying capacity of a stream or river constantly changes both spatially and temporally as flow changes. Flow in any river or stream is never homogenous, so the resulting sediment movement in any section of stream or river varies greatly.

Particles that are too heavy to be fully suspended may roll or slide along the bed (*traction load*) or hop as they rebound on impact with the bed. In the latter case, ballistic trajectories occur, and the particle is said to move by *saltation*. *Stream competence* refers to the heaviest particles that a stream can carry. Stream competence depends on stream velocity; the faster the current, the heavier the particle that can be carried. *Stream capacity* refers to the maximum among of total load (bed and suspended) a stream can carry. It depends on both discharge and velocity, since velocity affects the competence and therefore the range of particle sizes that can be transported. Note that as stream volume and discharge increase, so do competence and capacity. This is not a linear relationship, and doubling the discharge and velocity does not automatically double the competence and capacity. Stream competence varies as approximately the sixth power of velocity. For example, doubling velocity usually results in a 64 times increase in competence. For most streams, capacity varies as a range of squared to cubed values. For example, tripling the discharge usually results in a 9-27 times increase in capacity. Most of the work of streams is accomplished during floods, when stream velocity and discharge (and therefore competence and capacity) are many times their level compared to periods of quiescent flow. This work is in the form of bed sourcing (erosion), sediment transport (bed and suspended loads), and sediment deposition.

Suspended Solids and Turbidity

It has been stated that *total suspended solids* (TSS) in water are the most important pollutant. Erosion happens constantly around the planet, and some rivers and streams have naturally high TSS levels without any human intervention. The Yangtze River in China and the Colorado and Mississippi Rivers in the U.S. are examples of rivers that have historically entrained large amounts of sediment due to local topography, geology, and climate.

Total suspended solids are defined as all solids suspended in water that will not pass through a 2.0 μm glassfiber filter (dissolved solids would be the fraction that does pass through the same size filter). The filter is then dried in an oven between 103 and 105°C, and weighed. The increase in weight of the filter represents the amount of TSS.

Problems with TSS arise when excess erosion occurs in a watershed due to human land use practices. Excess levels of TSS can come from either point (municipal and industrial wastewater) or nonpoint (e.g., agriculture, timber harvesting, mining, and construction) sources. Generally, water with less than 20 mg/L is considered relatively "clear"; levels between 40 and 80 mg/L tend to be "cloudy"; while levels over 150 mg/L would be classified as "dirty" or "muddy." Point sources generally require treatment, usually through settling or flocculation, prior to being released into a river or stream. Nonpoint sources are much more difficult to manage due to several sources acting synergistically. No-till farming, sedimentation basins, and silt fences are common practices to reduce run off from agriculture or construction areas. Stormwater retention ponds and regular street sweeping can reduce the impact of stormwater runoff form urban areas.

Increasing levels of TSS often result in a waterbody being unable to support diversity of aquatic life. Sedimentation of the stream bed as velocity decreases often results in the suffocation of many aquatic macroinvertebrates and the eggs of fish. Where TSS is deposited results in increased *embeddedness* (the percentage of any piece of substrate covered in sediment) of cobble, rocks, and boulders within the stream. Several species of macroinvertebrates use the bottom of rocks as refuge from predators or from fast-flowing water, and as embeddedness increases, this vital habitat is diminished or completely lost. Additionally, TSS absorbs heat and can increase the temperature of a waterbody. In lakes and reservoirs, this can exacerbate thermal stratification as heat accumulates close to the surface. Suspended solids can also decrease the amount of dissolved oxygen due to consumption of organic matter by respiring bacteria. In lakes or reservoirs with a large algal biomass, sudden inputs of water containing suspended sediments have been known to deplete the oxygen of water and cause massive fish kills. The once-photosynthesizing algae switch to respiration as light for photosynthesis was reduced or eliminated.

Besides the relatively direct effects of suspended solids on water bodies, perhaps the greatest indirect effect are the pollutants that may be attached to suspended sediment. Examples of pollutants known to sorb to sediment particles are nutrients, metals, organic compounds such as polycyclic aromatic hydrocarbons (PAH) and polychlorinated biphenyls (PCBs), and a wide assortment of herbicides and pesticides. All of these contaminants have differing solubilities and therefore differing fates once they enter a river, stream, or lake. Sediment-associated pesticides in water are an emerging problem in agricultural areas.

Turbidity is related to, but not a proxy for, suspended sediments. Specifically, turbidity is the quantification of the light that is scattered or absorbed rather than transmitted through a water sample. Turbidity is another measure of water clarity but is not a measure of dissolved substances that can add color to water. Particulates are what add turbidity to water and can include such things as silt, clay, organic matter, algae and other microorganisms, and any other particulate matter that can scatter or absorb light. The amount of light scattered or absorbed is proportional to the concentration of particulates in the sample. The exact amount and wavelength of light scattered by a particle is dependent on the particle's shape, size, and refractive index, which makes any correlation between turbidity and suspended solids difficult and impractical. However, turbidity is directly related to the level of particulates and is an excellent general indicator of water quality in its own right. Units of measure for turbidity are in *nephelometric turbidity units* (NTUs) and are measured on a nephelometer (often called a turbidimeter). Turbidimeters operate by shining an intense beam of light up through the bottom of a glass tube containing the sample. Light scattered by particulates in the sample is detected by a sensitive photomultiplier tube at 90-degree angle from the incident beam of light. The amount of light reaching the photomultiplier tube is proportional to the level of turbidity in the sample. The photomultiplier tube converts the light energy into an electrical signal, which is amplified and displayed on the instrument meter.

Metals as Surface Water Contaminants

Metals that can be toxic to humans and wildlife are often found in industrial, municipal, and urban runoff and in atmospheric deposition from coal-burning plants and smelters and from natural weathering of rocks and soils. Levels of harmful metals in water have risen globally with increasing urbanization and industrialization. Currently, there are over 50 heavy metals that can be toxic to humans. Of these, 17 are considered very toxic and simultaneously readily accessible.

Common heavy metals known to be toxic to humans include arsenic, cadmium, chromium, copper, lead, mercury, and zinc. Interestingly, chromium, copper and zinc are essential micronutrients required by the human body for growth, and toxicity depends upon enhanced dose.

Heavy metals are also environmentally persistent, which exacerbates any potentially toxic exposure because these metals often accumulate under certain environmental conditions.

Mercury

Mercury is the environment is one of the most widely recognized and publicized pollutants. Under certain environmental conditions, elemental mercury complexes to form methyl-mercury, which is especially mobile in the environment and toxic to humans and wildlife. Use of mercury in the tanning industry and for making hats was the first time that it was widely recognized as a toxic substance affecting the brain; hence the term "mad as a hatter". Today, release of mercury in smokestack emission from coal-burning power plants is the primary source of contamination. Bioaccumulation of methyl-mercury in long-lived predatory fish from cold freshwater lakes (pikes and walleyes) and is cold water marine species (swordfish, sharks, and some tunas) has led to public warnings for pregnant and nursing women to limit consumption of shark and swordfish and for all of the public to limit consumption of fish form certain lakes in parts of the United States and Western Europe.

The cycling of mercury through the environment is complex and depends on several physical, chemical, and, most importantly, biological aspects of the system in question. The manner in which mercury cycles through any given area or ecosystem determines its relative toxicity and subsequent bioaccumulation rate upward through the food chain. The term *bioaccumulation* refers to the net accumulation, over time, of pollutants within an organisms from both biotic and abiotic factors. The term *biomagnification* refers to the progressive accumulation of persistent toxicants by successive trophic levels. Biomagnification relates to the concentration ratio in a tissue of predator organisms as compared to that in its prey. Mercury exists in several forms in the environment: elemental mercury (Hg^0), inorganic mercury compounds (Hg^{+1} or Hg^{+2}), and organic mercury compounds ($HgCH_3$ or $Hg(CH_3)_2$, which include both methyl- and dimethyl-mercury.

There are numerous pathways by which mercury can make its way into water bodies. Inorganic and methyl-mercury can enter water directly from atmospheric deposition, methyl-mercury and Hg^{+2} can be bound to organic substances in runoff, and surface water flow in upper soil layers can transport Hg^{+2} and methyl-mercury to water bodies. There has been a global increase of mercury released into the atmosphere since the beginning of the Industrial Age, so that atmospheric deposition onto watersheds and surface water often plays as large role as runoff from natural sources.

Once in an aquatic ecosystem, mercury goes through several complexation and transformation processes. While most forms of mercury are bioavailable methyl-mercury (MeHg) is the form most readily absorbed and bioaccumulated. The methylation of mercury in aquatic systems not only requires a certain range of physicochemical factors, but also the presence of a group of bacteria known as *sulfate-reducing bacteria* (SRBs). There are several species of SRBs, but some of the most common include strains of *Desulfovibrio* and *Desulfobacter*. The majority of methylation that occurs in lakes and reservoirs is within anaerobic sediments.

Table 5.1. Factors influencing the methylation of mercury in aquatic ecosystems

Physical or chemical condition	*Influence on methylation*
Low dissolved oxygen	Enhanced methylation
Decreased pH	Enhanced methylation within the water column
Decreased pH	Decreased methylation in sediment
Increased dissolved organic carbon	Enhanced methylation within water column
Increased salinity	Decreased methylation
Increased nutrient concentrations	Enhanced methylation
Increased temperature	Enhanced methylation
Increased sulfate concentrations	Enhanced methylation

Mercury in all forms is a potent toxin that can cause developmental effects in the fetus as well as toxic effects on the liver and kidneys of adults and children. Sublethal effects of mercury toxicity can affect the ability to learn, speak, feel, see, taste and move. Children under the age of 15 are most vulnerable, because their central nervous system is still developing, Mercury is easily passed form pregnant mother to fetus, and even extremely small trace amounts of mercury can have devastating effects to a developing central nervous system. Mercury toxicity can occur through skin contact, inhalation, or ingestion. Due

$$Hg^{0} \underset{\text{Reduction}}{\overset{\text{Oxidation}}{\rightleftarrows}} Hg^{+1} \text{or } Hg^{+2} \underset{\text{Demethylation}}{\overset{\text{Methylation}}{\rightleftarrows}} HgCH_3$$

Fig. 5.1. Common transformations of mercury.

to biomagnification, the most common route of exposure to humans is through consumption of contaminated fish. Approximately 60,000 babies are born in the U.S. each year with some degree of mercury toxicity.

The methylation, biomagnification, bioaccumulation, and toxicity of mercury are often linked to the problem of increasing eutrophication. Increases in most of the factors that cause eutrophication within a waterbody, including increased sulfate content, also increases the rate of methylation and subsequent toxicity to humans and wildlife. Humans have not only increased the availability of mercury deposited into aquatic systems, we have, through cultural eutrophication, also increased the bioavailability and subsequent toxicity of mercury in these systems.

Arsenic

Arsenic is an element widely distributed throughout the earth's crust. As such, it is often introduced into water through the dissolution of minerals and ores and may concentrate in groundwater. Arsenic is also used in industry and agriculture and is a by-product of copper smelting, mining, and coal burning. One form of arsenic, chromated copper arsenate, is the most common wood preservative in the U.S. and contains 22% arsenic.

Inorganic arsenic occurs in several different forms in the environment, but in natural waters it is most commonly found as trivalent arsenite [As(III)] or pentavalent arsenate [As(V)]. Most of the organic species of arsenic, usually at very high levels in seafood, are less toxic and are readily eliminated by normal body functions.

Symptoms from arsenic exposure include vomiting, esophageal and abdominal pain, and bloody diarrhea. Long term exposure can cause cancers of skin, lungs, urinary bladder, and kidney as well as other skin changes, such as pigmentation changes and thickening.

One of the largest mass poisonings in the world recently occurred in Bangladesh, where 53 out of a total of 64 districts had groundwater contaminated with arsenic. The cause of arsenic contamination was related to the onset of intense agriculture in the region where irrigation resulted in large-scale withdrawal of grounds via wells.

Chromium

Chromium is found in natural deposits as ores containing other elements. Additionally, chromium is an important industrial metal, where it is used in alloys such as stainless steel, protective coatings on other metals and magnetic tapes, pigments for paints, cement, paper, rubber, and floor coverings. Chromium has several oxidation states, but the most common are $^{+2}$, $^{+3}$ and $^{+6}$, with $^{+3}$ being the most stable. Oxidation states of $^{+4}$ and $^{+5}$ are relatively rare.

Toxicity of chromium depends on oxidation state. Chromium(III) is an essential nutrient, while the hexavalent form, chromium(VI), is believed to be carcinogenic in humans. Evidence to date indicates that the carcinogenicity is site-specific, limited to the lung and sinonasal cavity, and dependent on the high exposures.

Selenium

Selenium occurs naturally, in the environment as selenide and is often combined with sulfide copper, lead, nickel, or silver. Like chromium, selenium is a micronutrient needed in very small quantities in humans and wildlife to produce the amino acid selenocysteine. However, it can be toxic at higher doses. The relatively narrow range between selenium acting as a beneficial nutrient (50 μg/day) and the initiation of toxicity (400 μg/day) in humans means that it needs to be closely monitored in the environment, especially in areas with alkaline soils, because this is where selenium is often found in its most oxidized and toxic form. As with several other naturally occurring heavy metals, problems may arise due to increased availability of selenium in aquatic systems primarily due to irrigation and farming practices. Symptoms of short-term selenium toxicity include hair and fingernail changes, damage to the peripheral nervous system, and irritability. Long term symptoms include damage to liver and kidney tissue and nervous and circulatory systems.

Selenium is a bioaccumulative pollutant; however, unlike mercury, selenium concentrations do not increase upward through the food chain, i.e., it does not biomagnify. Selenium toxicity can have devastating effects on both terrestrial and aquatic wildlife. Selenium can affect the growth and survival of juvenile fish as well as the offspring of adult fish exposed to sublethal levels. Birds that have eaten fish suffering form selenium toxicity either succumb to the acutely toxic effects of selenium or produce offspring, often stillborn, with gross skeletal deformities. Due to selenium uptake in terrestrial plants, both domestic and wildlife species foraging on these plants can be affected.

Nutrients and Eutrophication of Surface Waters

On a global scale, eutrophication has often been cited as the number one can cause of impairment to surface water resources.

Eutrophication is the gradual accumulation of nutrients, and organic material subsequently utilizing these nutrients as an energy source, within a body of water. While eutrophication is often cited as an example of anthropogenic pollution of inland waters such as lakes and streams; coastal areas, estuaries, and salt marshes are also commonly affected. Eutrophication often results in increases in algal biomass, and therefore some discussion of what these nutrients are, and what specific ratios cause eutrophication, are in order.

Justus Von Liebig, a German analytical chemist and professor of chemistry at the University of Giessen, made great contributions to the science of plant nutrition and soil fertility in the mid-1800s. Liebig's *Law of the Minimum* states that yield is proportional to the amount of the most limiting nutrient, whichever nutrient it may be. From this, it may be inferred that if the deficient nutrient is supplied, yields may be improved to the point that some other nutrient is needed in greater quantity than the soil can provide, and the Law of the Minimum would apply in turn to that nutrient. This same law can be applied to aquatic system. The nutrients that most often limit primary production in aquatic systems are forms of carbon, nitrogen, and phosphorus. The specific ratio of limitation (on a molar basis) is 106C:16N:1P. *Carbon* is ubiquitous in the environmental and atmosphere, but can become limiting during in tense photosynthesis by algae or aquatic plants. Since carbon dioxide is utilized during photosynthesis, it is possible for this carbon nutrient to be temporarily depleted during daylight hours. This situation would be reversed during the evening, when respiration would exceed photosynthesis, and the carbon dioxide utilized during the day would be released back into the water. Generally, it is uncommon for carbon to be a limiting nutrient.

The idea of nutrient limitation based upon the ratio between C:N:P in aquatic systems only works when one of these essential nutrients is, indeed, "limiting" the growth of primary producers such as algae. In eutrophic or hypereutrophic systems, ratios may indicate that a nutrient is "limiting" in the traditional sense, however, if all nutrients are orders of magnitude higher than what it takes to limit primary production than, in this case, ratios can be misleading and nothing is truly limiting growth. Often, in hypereutrophic systems, algal biomass can become so large that the only limiting factor is available light for photosynthesis, as algal cells near the surface shade those at depth.

Nitrogen is an essential plant nutrient used in the synthesis of organic molecules such as amino acids, proteins, and nucleic acids.

Nitrogen (mostly as N_2 or "dinitrogen" gas) comprises 78% of the Earth's atmosphere. Most of the abundant nitrogen must be "fixed" into nitrate (NO_3^-), ammonia (NH_3), or ammonium (NH_4) before it can be used by organisms incapable of fixation. Organisms capable of nitrogen fixation include certain species of bacteria, actinomycetes, and cyanobacteria. In aquatic systems, cyanobacteria perform the majority of nitrogen fixation. In aquatic systems, the forms of nitrogen of greatest interest are (in order of decreasing oxidation state):

-Nitrate (NO_3)

-Nitrite (NO_2^-)

-Ammonia (NH_3)

-Organic-N (amino groups)

Total oxidized nitrogen is the sum of NO_3^- + NO_2^-. Organic nitrogen is the organically bound fraction and includes such natural materials as proteins and peptides, nucleic acids and urea, and numerous synthetic organic materials. Analytically, organic nitrogen and ammonia can be determined together and referred to as "Kjeldahl nitrogen", a term that reflects the technique used in their determination. Total Kjeldahl nitrogen (TKN) is not synonymous with total nitrogen. If TKN and NH_3 are determined individually, "organic nitrogen" can be estimated by the difference.

$$\text{TKN} + NO_3^- + NO_2^- = \text{Total nitrogen}$$

All forms of nitrogen (organic and inorganic) are interconvertible. The nitrogen cycle is an important component of overall biogeochemical cycling in aquatic systems.

Ammonification is an important process in the nitrogen cycle and, is basically, the process of decomposition with production of ammonia or ammonium compounds especially by the action of bacteria on organic matter. Aquatic animals commonly excrete NH_3 as a waste product of metabolism. The excreted or mineralized NH_3/NH_4 is then available for direct uptake and utilization by other organisms, or it may be converted to more oxidized forms of nitrogen for incorporation into cells. In some nitrogen-poor lakes or reservoirs, the excretory contribution (eg. ammonification) from zooplankton can provide up to 90% of the nitrogen required by primary producers. Ammonification is difficult to quantify because of the rapid uptake of NH_3 and NH_4 by primary producers. Ammonification is the opposite of assimilation and protein synthesis. Both aerobic and anaerobic bacteria play vital roles in ammonification.

Nitrification is the biological oxidation of NH_4^+ and NH_3 to NO_2^- and then NO_3. Nitrification is an important because NH_4^+ and NH_3 are toxic to species of aquatic vertebrates. Nitrification is performed by bacteria that gain energy form oxidizing reduced forms of nitrogen. The aerobic chemoautotrophs involved in nitrification are species of *Nitrosomonas* and *Nitrobacter*. Nitrification consumes and simultaneously requires oxygen, and is a two-part process

$$NH_4^+ + 1/2\,O_2 \xrightarrow{\text{Ammonia mono-oxygenase}} NH_2OH + H^+$$

$$NH_2OH + O_2 \longrightarrow O_2^- + HOH + H^+$$

This process requires 66 Kcal of energy/gram atom of ammonium oxidized.

Under anaerobic conditions:

$$NH_2OH \rightarrow NOH \rightarrow N_2O$$

Ammonium oxidation has important ecological significance in aquatic systems. This microbes that perform nitrification are relatively inefficient autotrophs that use the energy gained form oxidizing ammonia to fix carbon. Thus, these bacteria have a dual ecological role: they are involved in recycling nitrogen and in fixing carbon into organics. The microbes that perform nitrification are fragile. These organisms are acid-sensitive even though they produce acid. If a large source of nitrogen is added into the environment, these organisms can potentially kill themselves by metabolizing it to nitric acid. Since they are also strict aerobes, they can be e killed if introduction of wastes leads to excessive growth of other species that deplete oxygen (i.e., eutrophication).

Denitrification is the reduction of nitrate (NO_3) to nitrogen compounds and can be a significant pathway for the loss of nitrogen from aquatic systems. There are two types of denitrification, assimilatory and dissimilatory.

Assimilatory nitrate reduction. Many organisms can only acquire nitrogen in the form of nitrate and must reduce nitrate to form the amino groups needed for metabolism.

$$NO_3^- + \text{energy} \rightarrow \text{amino groups}$$

The "energy" in the above equation is usually supplied by enzymatic activity (nitrogenase).

Dissimilatory nitrate reduction. Dissimilatory nitrate reduction is performed by anaerobic bacteria that use nitrate as the terminal electron acceptor in the absence of oxygen. The overall equation is:

$$NO_3^- \rightarrow NO_2^- \rightarrow NO \rightarrow N_2O \rightarrow N_2 \text{ gas}$$

The individual steps of dissimilatory nitrate reduction are:

1. Reduction of nitrate to nitrite

   $2HNO_3^- \rightarrow 2HNO_2^- + 4e$

   Enzyme: dissimilatory nitrate reductase

2. Reduction of nitrite to nitric oxide

   $2HNO_2^- \rightarrow 2NO + 2e$

   Enzyme: dissimilatory nitrate reductase

3. Reduction of nitric oxide to nitrous oxide

   $2NO \rightarrow N_2O + 2e$

4. Reduction of nitrous oxide to dinitrogen

   $N_2O \rightarrow N_2 + 2e$

   Enzyme: dissimilatory nitrous oxide reductase.

Since reductions are energy yielding, 24 ATPs are generated per mole of nitrate reduced.

Although denitrification requires anoxic conditions, it has been observed in aerated lake sediments and can form relatively thin biofilms on rocks in streams. Evidently, denitrification can occur in microzones of anoxia within sediments and biofilms. Oxygen produced through photosynthesis by benthic algae may inhibit denitrification. Denitrification requires an organic carbon source and proceeds faster where more carbon is available in the water and sediments. Denitrification may contribute significant portion of the oxidative metabolism in water bodies where nitrate levels are high. Within any given waterbody, denitrification can occur simultaneously with nitrification. Denitrification occurs due to microorganisms, usually facultative anaerobes and predominantly two genera: *Pseudomonas* and *Bacillus*. Dissimilatory denitrification is used in sewage treatment and bioremediation where denitrifying bacteria aid in converting organic nitrogen to nitrogen gas that escapes to the atmosphere.

As explained by the previous processes, all forms of nitrogen and interconvertible. While there are losses of nitrogen within any aquatic system, there are simultaneous gains from the atmosphere and from recycling within any given region. Problems with eutrophication arise when humans contribute to loading of nitrogen to a waterbody form either point or nonpoint sources of pollution.

Since the 1940s, the amount of nitrogen available for uptake in aquatic systems at any given time has more than doubled. Human activities now contribute more to the global supply of fixed nitrogen each year than natural processes. Anthropogenic nitrogen totals about

210 million metric tons per year, while natural processes contribute about 140 million metric tons. This influx of extra nitrogen has caused serious distortion of natural nutrient cycling in aquatic systems. Excess nitrogen can wreak havoc with aquatic ecosystem structure affecting the number and kind of species found.

Phosphorus, like nitrogen, is essential to all life. Phosphorus functions in the storage and transfer of a cell's energy and in genetic systems. Cells use adenosine triphosphate (ATP) as an energy carrier that drives a number of biological processes, including photosynthesis, muscle contraction, and the synthesis of proteins. Phosphate groups are also found in nucleotides and therefore nucleic acids. Phosphorus is usually more scarce environmentally than other principle atoms of living organisms including carbon, hydrogen, oxygen, nitrogen, and sulfur.

Phosphorus occurs naturally in rocks and other mineral deposits. During weathering, the rocks gradually release the phosphorus as phosphate ions, which are soluble in water, and the mineralized phosphate compounds breakdown. Phosphorus exists primarily as phosphates in two forms: orthophosphate and organically bound phosphate. These forms of phosphate occurs in living and decaying plant and animal remains as free ions, chemically bonded, or mineralized and chemically bonded in sediments. Analytically, phosphorus in water is usually categorized as being either dissolved or particulate, depending on whether or not it can pass through a 0.45-μm filter. The "dissolved" fraction can have a substantial colloidal component. Within the dissolved fraction, inorganic P (dissolved inorganic phosphorus) occurs as orthophosphate (PO_4). Dissolved inorganic phosphorus is sometimes referred to as *soluble reactive phosphorus* (SRP). *Total phosphorus* (TP) is determined on a nonfiltered sample by heat and acid digestion, which converts the sample to SRP for measurement.

In unpolluted rivers, SRP averages about 0.01 mg/L on a worldwide basis and total phosphorus averages about 0.025 mg/L. Agricultural activities may increase SRP levels to 0.5-0.1 mg/L, and municipal effluents may increase SRP concentrations to 1.0 mg/L or much higher. Particulate phosphorous includes P incorporated into mineral structures, absorbed onto clays, and incorporated into organic matter. Worldwide averages of particulate phosphorous concentrations are about 0.5 mg/L. This level can be much higher depending upon land use and erodibility of the watershed.

Phosphorous is often the limiting macronutrient with regard to primary production in aquatic systems. Because of this, and its relative scarcity, it is quickly removed from its dissolved state and incorporated into living biomass. Bacteria and algae are both responsible for turnover rates as fast as 1–8 minutes. Turnover rates usually follow the order of (in order to decreasing turnover times): Bacteria → algae → zooplankton → vertebrates.

It's been estimated that in freshwater lakes, zooplankton excrete about 20% of the phosphorous required by phytoplankton, whereas bacteria can excrete up to 80%. Therefore, food web dynamics play a large role in either the sequestration or the recycling of phosphorus in aquatic systems. The speed at which phosphorous is moved between biotic and abiotic compartments makes interpretation of different forms difficult. It's impossible to distinguish between zooplankton-P, bacterial-P, algae-P, and sometimes even inorganic-P. The best way to quantify phosphorous in a body of water is by analysis of total phosphorous.

Eutrophication, besides increasing algal biomass, often results in depletion of dissolved oxygen, increases in pathogenic bacteria and viruses, increases in potentially toxic species of algae, fish kills and loss of biodiversity. Remediation efforts of even a small lake are usually cost-prohibitive, and it is very difficult, if not often impossible, to return a lake or reservoir back to an earlier trophic state. The best approach is a protective, watershed-based one that attempts to protect waterbodies from cultural eutrophication. This usually requires collaboration among several resource agencies, in addition to municipalities and individual landowners in the watershed.

Harmful Algal Blooms

Planktonic (i.e., free-floating algae) are vitally important components of all marine and freshwater systems on the planet. They form the base of the food chain in all aquatic systems. Of the thousands of known species, a few hundred have the potential to produce a wide variety of toxins under certain environmental conditions. Eutrophication greatly exacerbates the growth and prevalence of potentially toxic species.

Harmful algal blooms in marine systems

Most *harmful algal blooms* (HABs) occur in coastal areas where terrestrial runoff of nutrients cause the growth and proliferation of sometimes monospecific blooms of toxic algae. Dinoflagellates (Division Dinoflagellata) are marine phytoplankton often associated with toxic blooms.

Dinoflagellates are microscopic, unicellular, flagellated protists that can be either autotrophic (photosynthetic) or heterotrophic (consuming outer organisms). Heterotrophic forms often have life history patterns more akin to an animal than a plant. Additionally, dinoflagellates are routinely found in freshwater and often produce some of the same toxins found in marine systems.

Dinoflagellates have the potential to produce a variety of toxins that can be harmful to humans and wildlife. Some affect humans following the ingestion of shellfish or fish that have consumed toxic dinoflagellate species.

Ciguatera poisoning is the most commonly reported disease associated with consumption of seafood. Ciguatera is a lipid-soluble toxin that can affect a variety of fish species and can be very toxic to humans after ingestion of these fish. Tropical and subtropical fish species, including barracuda, grouper, and snapper, are commonly affected. The dinoflagellate species most often associated with ciguatera poisoning is *Gambierdiscus toxicus* but other species including *Prorocentrum mexicanum*, *P. concavum*, *P. lima*, and *Osteropsis lenticularis* have also been implicated. Ciguatera exhibits both gastrointestinal and neurological symptoms, with the time to onset usually less than 24 hours. Gastrointestinal symptoms include diarrhea, abdominal pain, nausea and vomiting. The most common neurological symptoms include abnormal or impaired skin sensations, vertigo, lack of muscle coordination, cold-to-hot sensory reversal, myalgia (muscular pain), and itching. Neurological symptoms may recur intermittently, with gradually diminishing severity for a long as six months. No deaths have been reported from ciguatera in the U.S., although worldwide, the mortality rate is 7-20% of people infected.

Brevetoxin is a large, lipophilic, polyether toxin primarily produced by the dinoflagellate Karenia brevis. Brevetoxin poisoning occurs with the most frequency in the Gulf of Mexico and has caused sporadic fish kills for decades. These toxins also affect shellfish, which can in turn poison humans who ingest contaminated shellfish. This syndrome is referred to as *neurotoxic shellfish poisoning* (NSP) and produces similar symptoms similar to ciguatera poisoning. There have been no reported fatalities from NSP, although it has been known to kill laboratory mammals.

Under certain environmental conditions, dinoflagellates can rapidly multiply in numbers and form "tides". Usually, tides are identified by the colour of the dinoflagellate causing the bloom. "Red" tides are often associated with species of *Alexandrium* and "brown" tides with species of *Aureococcus*.

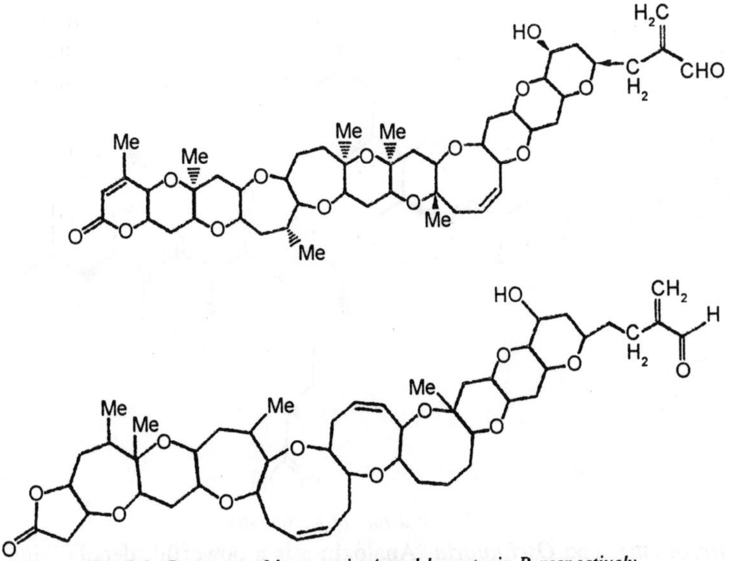

Fig. 5.2. Structure of brevetoxin-A and brevetoxin-B respectively.

This initiation of either red or brown tides is complex due to the complex life cycles of dinoflagellates but usually involves warm, nutrient-enriched water.

Harmful algal blooms in freshwater systems

While freshwaters often contain many of the same dinoflagellate species found in marine systems, and sometimes the same toxins, the majority of freshwater toxins are caused by several different species of cyanobacteria. Cyanobacteria, like dinoflagellates in marine systems, greatly increase in number in eutrophic waters and occur on a global scale. Cyanobacterial toxins ("*cyanotoxins*") can affect both humans and wildlife. Humans are usually affected from ingesting water containing cyanotoxins, and the disease is categorized based upon the type of toxin in the water. Toxins produced by cyanobacteria can be either hepatotoxic or neurotoxic.

One of the ubiquitous cyanotoxins is microcystin, which can be produced by species of *Anabaena*, *Nodularia*, *Nostoc*, *Oscillatoria*, and *Microcystis*. There are over 50 different analogues of microcystin. These toxins mediate toxicity by inhibiting liver function (i.e., *hepatotoxic*) and can often be found at high levels in drinking water reservoirs.

Anatoxin-a is a small, low -molecular-weight neurotoxic alkaloid produced by species of *Anabaena*, *Aphanizomenon*, *Cylindrospermum*,

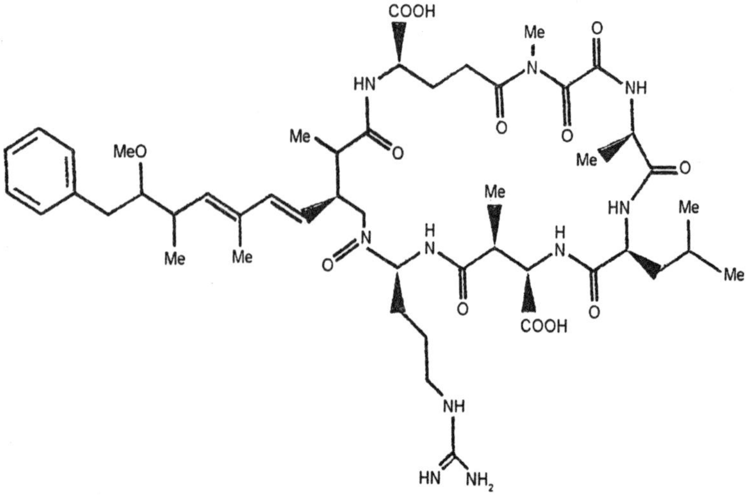

Fig. 5.3. Structure of microcystin.

Microcystis, and *Oscillatoria*. Anatoxin-a is a powerful, depolarizing, neuromuscular blocking agent that strongly binds to the nicotinic acetylcholine receptor. This is a potent neurotoxin that can cause rapid death in mammals through respiratory arrest.

Cylindrospermopsin, while having many of the properties of hepatotoxin, also resembles neurotoxin. It is produced primarily by *Cylindrospermopsis raciborskii*, but has also been found in *Umezakia natans* and *Aphanizomenon ovalisporum*. Cylindrospermopsin has poisoned at least 149 people, many of them children requiring hospitalization, in Palm Island, Queensland, Australia. At one time believed to be strictly a tropical to subtropical species, *C. raciborskii* has been found in waters in the north temperate U.S. recently. Like most cyanotoxins, cylindropsermopsin can often be found in drinking water reservoirs.

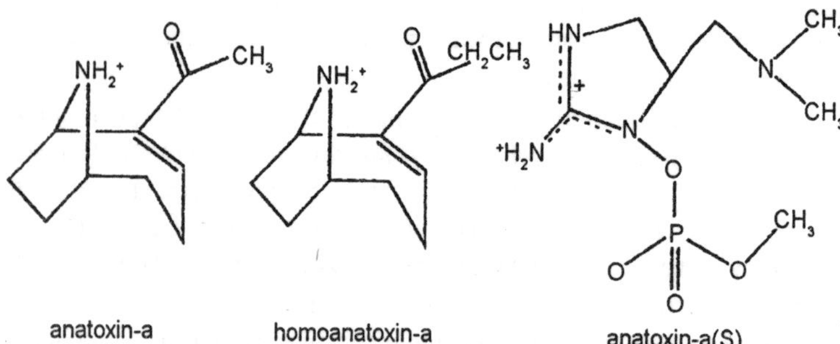

Fig. 5.4. Structure of anatoxin-a and the homologue homoanatoxin-a and anatoxin-a(s).

Organic Compounds in Water

Persistent, Bioaccumulative Organic Compounds

Certain organic compounds, due to their physiochemical properties, are very persistent in the environment. These compounds are referred to as persistent, bioaccumulative, toxic (PBTs) contaminants. These compounds are the most important organic contaminants in aquatic systems. These compounds bioaccumulate and biomagnify within the aquatic ecosystem, and they often accumulate in the sediments of surface water bodies.

Dichlorophenyltrichloroethane (DDT) is an organochlorine pesticide that was in widespread use from the 1940s to the early 1970s, when it was used primarily for agricultural crops or vector-pest control (the control of insects known to carry malaria and typhus).

When DDT is released into the environment, it begins to degrade into several different metabolites. Once bound to sediment particles, DDT and its degradation products can persist for up to 15 years depending upon environmental conditions. Water is the main route of exposure of DDT and its metabolites to humans and wildlife. Ingestion of foodstuffs and in particular consumption of fish is how humans ingest the largest amount of DDT, primarily due to bioaccumulation. In fish and other wildlife, especially predatory birds feeding on fish, even if acute toxicity and death does not occur, reproductive failure often results.

The use of DDT in the US has been banned since 1972, however, the need to protect agricultural crops and humans from insect-borne vectors of disease still exists. Most organochlorine pesticides, including DDT, have been replaced with less environmentally persistent compound such as organophosphate, carbamate, and synthetic pyrethyroid pesticides. While these compounds are degraded in the environment at a much faster rate than DDT, they are also more acutely toxic. Even though DDT was banned over 30 years ago, due to its persistence, we still feel its toxic effects in the U.S. Twenty years after the ban of DDT, the U.S. EPA reported that out of 388 sites throughout the nation sampled between 1986 and 1989, total DDT and PCBs were detected at 98 and 90% of all sites respectively. Fish still remain vulnerable to the effects of DDT. A study of Munn and Gruber in 1997 showed total DDT was detected in 94% of whole-fish samples collected in streams of eastern Washington state.

Polychlorinated biphenyls (PCBs) are a group of organic compounds with similar physical structure and chemistry, ranging form oily liquids

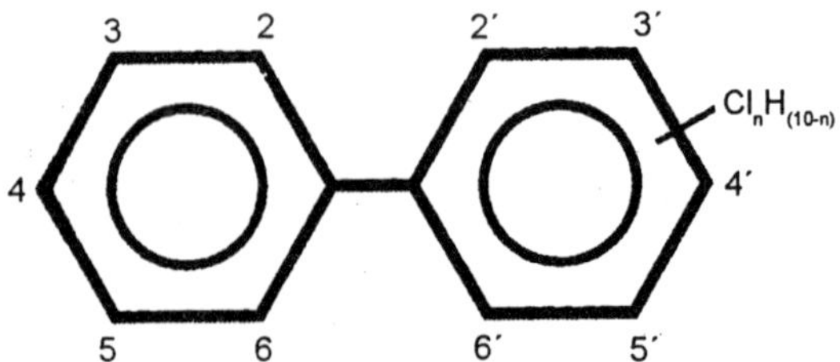

Fig. 5.5. Basic structure of a polychlorinated biphenyl.

to waxy solids. All PCBs are formed from the addition of chlorine (Cl_2) to biphenyl ($C_{12}H_{10}$), which is a dual-ring structure consisting of two-carbon benzene rings linked by a single carbon-carbon bond. The presence of a benzene ring allows a single attachment to each carbon, meaning that there are 10 possible positions for chlorine to replace the hydrogens in the original biphenyl.

Any unique compound in the PCB category is referred to as a "congener" whose individual name is dependent upon the total number and position of each chlorine substitute. There are 209 PCB congeners.

Due to the chemical stability and high boiling point of PCBs, they were used to hundreds of industrial applications, including electrical insulation, hydraulic equipment, and plasticizers in paints, plastics, and rubber products. Prior to their ban in 1977, total production of PCB in the U.S. was more than 1.5 billion pounds. Because of the vast amount of possible congeners, PCBs were sold as many different trade names but one of the most prevalent was Arachlor.

Similar to DDT, PCBs are environmentally persistent and adhere strongly to particulates in water, meaning that they can remain intact in sediments of lakes and rivers for extended periods. Because of PCBs' strong adherence to sediments and suspended particles in water, actual contamination of a waterbody may be several times higher than the actual solubility of a particular PCB. Also like DDT, all PCBs are extremely lipophilic, meaning that they can bioaccumulate and biomagnify in aquatic environments. First detected in the 1960s, PCBs were found to be contaminants on a global scale, occupying virtually every component of the environment including air, water, soil, fish, wildlife, and human blood.

Uptake of PCBs by microorganisms is very rapid and extremely high bioconcentration factors are often seen. Uptake by microorganisms is by true absorption into cells rather than adsorption onto the cell.

Fish are especially susceptible to the accumulation and concentration of PCBs, and all life stages of almost every species readily absorb

PCBs from the water. PCB congeners with higher chlorination levels are taken up most rapidly by fish. Due to the fact that PCBs are usually at higher levels in sediments, fish such as bottom feeders are most susceptible; however, route of exposure in fish can occur through water, sediment, or prey. Due to rapid uptake of PCBs in fish tissue, birds, especially those that eat fish, are also vulnerable. Egg-laying females can transfer substantial amounts of PCB to eggs with subsequent reproductive failure.

Route of exposure to humans, like DDT, is generally much greater for aqueous environments (through either direct ingestion of water or through eating contaminated fish) than terrestrial. PCBSs probable carcinogens in humans and are known to be carcinogenic to laboratory animals. The risks associated with consuming fish contaminated with PCBs are more than 1,000 times greater than the 1-in-a-million cancer risk used to regulate most hazardous wastes.

ENTER PATHOGENS AS SURFACE WATER CONTAMINANTS

Almost all animals are capable of excreting disease-causing intestinal microorganisms (enteric pathogens) in their feces. Sources of pathogens into surface waters include:

1. Urban storm water
2. Combined sewer and sanitary sewer overflows
3. Animal feeding operations
4. Sewage treatment plants
5. Septic tanks (onsite systems)

Pathogens can remain infectious for prolonged periods of times in surface waters presenting health risks to recreational users shellfish harvesting, and drinking water treatment plants. While drinking water treatment plants are required to treat water from surface sources, the more pathogens that are present in the raw water, the more treatment is required. Also, after periods of heavy rains, when the amount of suspended matter and pathogens often increases, it is difficult to remove all pathogens. For example, it has been shown that waterborne disease outbreaks in the United States are related to the intensity of rainfall events.

Forty percent of rivers and estuaries that fail to meet ambient water quality standards fail because of pathogens, usually measured by fecal coliform bacteria.

Stormwater can contain a wide variety of pathogens that originate form the feces of wild and domestic animals. Besides pets, other animal sources in urban areas include pigeons, geese, rats, and raccoons. Animals feces accumulate on the ground, and following a storm event are flushed into nearby streams and lakes. This results in a rapid increase in the concentration of enteric organisms, sometimes exceeding that found in raw sewage. In some cites, sewers that collect domestic sewage are combined with stormwater drains or collection systems. These flows are then transported to a sewage treatment plant for treatment. Unfortunately, after periods of heavy rainfall, this combined flow is greater than the sewage plant can treat, requiring the sewage plant to discharge untreated combined sewage and stormwater. These events are referred to as *combined sewer overflows* or CSOs. CSOs generally occur in order parts of the country, involving approximately 900 cities. To reduce the impacts of CSOs, cities may blend the untreated wastewater with treated wastewater or may construct large holding reservoirs where the combined flows can be stored until they can be treated later.

In the United States, there are 238,000 animal feeding operations, or AFOs (sometimes called confined animal feeding operations, or CAFOs) that produce 350 million tons of manure annually. This figure does not include manure from grazing animals. These operations generate approximately 100 times as much manure as municipal wastewater treatment plants produce sewage sludge (biosolids) in the United States. The Clean Water Act requires operations having more than 1,000 animals to have a discharge permit. Runoff form AFOs and farmland can contribute significant levels of pathogens that can infect humans, such as *Cryptosporidium* and *Escherichia coli*.

Septic tanks (also referred to as *decentralized* or *onsite wastewater treatment systems*) collect, treat and release about 4 billion gallons of treated effluent per day from an estimated 26 million homes, business, and recreational facilities in the United States. Poorly treated wastewater from improperly operating or overloaded systems can contain enteric pathogens that by transport through the soil can make their way to nearby streams can result in the sewage reaching the surface. The discharge of partially treated sewage from malfunctioning onsite systems was identified as a principal or contributing source of degradation in 32% of all harvest-limited shellfish growing areas in the United States. Problems with surface water contamination by onsite systems are most likely to occur in areas with shallow groundwater tables (e.g., within a few feet of the surface).

In the United States it is required that sewage discharges be disinfected to reduce the level of pathogens. While very effective in eliminating most enteric bacterial pathogens, significant levels of enteric virus and protozoan parasites may remain.

Total Maximum Daily Loads (TMDL)

A *total maximum daily load* (TMDL) is the maximum amount of pollution that a waterbody can assimilate without violating water quality standards. A TMDL is the sum of the allowable loads of a single pollutant from all contributing point and nonpoint sources, so that a waterbody can meet a designated use, such as swimming or fishing. Under the Clean Water Act of 1972, states are required to identify surface waters not meeting water quality standards and develop a TMDL for each pollutant for each listed waterbody. Once the impaired body of water is identified, a study is usually conducted to identify the sources and concentration of pollutants. From this, an informational plan is developed to reduce the most significant source(s) so that water quality standards can be met.

TMDL = point sources of wastes allocations (e.g., sewage treatment plant discharge) + non-point source load allocations (urban runoff + natural sources (mineral deposits, wild animals) + growth factor (growth of enteric bacteria) + a margin of safety (to compensate for uncertainties about the link between pollutant loads and impairments)

A load allocation is the part of a TMDL/water quality restoration plan that assigns reductions to meet identified water quality targets. For example, in a given watershed it was found that cattle were the major source of fecal coliform bacteria in the streams. To reduce fecal coliform loading, fences were placed to limit direct access of the cattle to the stream.

Fecal coliform bacteria and temperature (largely thermal waters from power plants) are the major contaminants that cause most waterbodies to not meet water quality standards for intended uses in the Untied States.

Quantification of Surface Water Pollution

Even as treatment methods were being developed, methods were simultaneously developed to quantify and assess both the disease threat and the dissolved oxygen problem posed by the discharge of municipal wastes into waterbodies. Quantitative methods are also used to calculate the degree of treatment needed. Such methods are base don an understanding of the physical and biochemical processes controlling the decay of microbes and chemicals over time.

Die-Off of Indicator Organisms

Early in the development of quantitative assessment methods, scientists realized that the many different microorganisms that exist in human wastes could not be effectively cultured and counted. Consequently, they settled on the coliform group of organisms to serve as an indicator of fecal pollution because they could be cultured and counted easily. Known to exist in large numbers in the gut of all warm-blooded animals, the coliform group provides a good indication of fecal pollution: however, it is not very specific, so other indicators such as fecal coliforms and streptococci may also serve as indicators of the sanitary quality of water.

Tests have shown that 99.99% of the indicator bacteria can be removed by wastewater treatment. But the residual 0.01% remains a problem raising concerns about the quality of water for recreational use, for example. The fact is that the number of bacteria in sewage is tremendous (500 million to 2 billion per 100 milliliters), so that, in assessing quality, the percentage of removal is not as useful as is the actual remaining concentration. The microorganisms concentrations allowed for various uses of water are relatively small. For drinking purposes, the concentration of fecal coliforms should,, of course, be 0, but a concentration of less than 1 per 100 mL may be allowed; for bathing, a concentration of 100 per 100 mL is frequently accepted. If, for purposes of demonstration, we assume that raw sewage has a population of one billion (10^9) per 100 mL, a removal efficiency of 99.99% would still leave 100,000 per 100 ml. The good news is that the concentrations of organisms decreases with time and distance downstream of the discharge point owing to natural processes. Depending on the location and proximity of uses, the point of discharge and method of discharge can be designed to optimize the rate of natural purification, further reducing downstream pollution problems. The bad news is that we have only a partial understanding of all the processes that affect die-off.

The concentration C of bacterial indicators of fecal contamination has been observed to decrease with time t according to a first-order reaction, the equation of which is

$$\frac{dC}{dt} = -KC \qquad \ldots(1)$$

where K is the *die off rate constant*. One parameter frequency employed in water pollution analyses obtained by solution of the first-order reaction equation is t_{90}, which is the time required for 90% die-off of the bacteria. This parameter is analogous to the half-life (t_{50}) used in radioactivity studies and is calculated in the same fashion. Thus, given the value of K

$$t_{90} = \frac{2.3}{K} \qquad ...(2)$$

Thee general solution of Equation (1) is used to find the concentration at any time C_t, after the initial concentration C_i is determined:

$$C_t = C_i e^{-Kt} \qquad ...(3)$$

This information can be used to calculate freshwater or marine die-off. To find the concentration of bacterial after effluent has traveled in the river for, say, 8 hours, it is necessary to determine the value of K for the specific situation being studied. Results of many studies in lakes and streams have shown that K varies widely, depending on the temperature of the water, the amount of sunlight, and the depth at which the plume travels. An average value for fresh water is about $K = 0.038$ per hour; however, values from 0.02 to 0.12 per hour have been measured. Using Equations (3) and $K = 0.038$ per hour, the initial concentration $C_0 = 100{,}000$ per 100 mL would be reduced to about 74,000 per 100 mL in $t = 8$ hours due to die-off alone. In many cases, it is preferable to predict the concentration at a given distance downstream, rather than its time increments. Most U.S. rivers have a remarkably uniform low-flow travel time to distance. In our example, this travel distance would be between 12 and 16 km. Travel times for other flow conditions vary from river to river, so field measurements may be needed to relate bacterial concentrations to specific locations downstream.

In analyzing coliform die-off cases in the marine environment, we often use an average value of 1.2 per hour for K, which is nearly 30 times greater than that of freshwater. This rapid die-off rate is usually attributed to the salinity of the marine environment, although it may also be related to a greater concentration of predatory animals. In addition, the natural flocculation and sedimentation of particles that occurs in estuaries could account for removal of bacterial from the water column. There are, however, other factors that can reduce, the die-off rate. For example, when an effluent is discharged at a great depth, the die-off can slow down considerably because sunlight cannot penetrate deeply enough.

Calculating seawater die-off is similar to freshwater die-off. The value of K for marine waters usually ranges from 0.3 to 3.8 per hour. Recently, however, K values as low as 0.02 per hour have been found where an effluent plume is transported in a layer far below the surface, say, 40 meters, suggesting that die-off is reduced because of the low penetration of UV radiation to that depth. What difference would this low K value make in the concentration?

Using the average value of $K = 1.2$ per hour, the original concentration, $C_0 = 100{,}000$ per mL, would die off to $C_1 = 6.7$ per 100 mL in $t = 9$ hours using Equation (3). But using $K = 0.02$ per hour instead of 1.2 per hour, we get 85,000 per 100 mL. Thus, for a 60-fold reduction in K, the concentration is increased by a factor of 85,000/6.7 = 13,000. It is evident from this example how important it is to have accurate values for K and how widely the results can vary with equally good, but different, estimates for K.

Variations in the rates of indicator bacterial die-off are not the only problem we have to contend with. Noncoliform pathogenic microorganisms may decay at rates different form those of our coliform indicators. Therefore, as water-analysis techniques become more sophisticated, we will need to conduct many field observations to establish values that can be used to predict die-off rates for specific pathogens.

In addition to the decrease of bacterial concentrations due to die-off in either fresh or marine surface waters, bacterial concentrations are decreased as the water is diluted with upstream ambient water at the point of effluent discharge and further diluted as it flows downstream. The effect of dilution may or may not be important in meeting water quality criteria, depending on the initial mixing, the nature of the subsequent flow patterns, and the distance to water use areas. But before discussing the mechanics of the dilution process, observe how the same first-order decay process used for assessing indicator bacteria die-off can be applied to the analysis of the fate of biodegradable organics.

Organic Matter and Dissolved Oxygen

Biodegradable organic compounds are decomposed by bacteria and other organisms that live in surface waters. While some organics are mineralized to carbon dioxide and oxides of nitrogen, others are synthesized into more microbial biomass, most of which is subsequently decomposed as well. All this decomposition consumes dissolved oxygen (DO), upon which many desirable species of fish, other aquatic organisms, and wildlife depend. Thus depressed dissolved oxygen concentrations adversely affect these life forms. For example, some fish can survive at concentrations near 1 mg L^{-1}, most are adversely affected at Do concentrations below 4 mg L^{-1}. The maximum amount of oxygen that pure surface water can hold is a function of salinity, temperature, and atmospheric pressure; compared to the maxima of many other substances, however, it is remarkably low. Note that solubility of O_2 *decreases* with increasing temperature, which is the opposite of the temperature-solubility relationship observed for most substances in water.

Domestic sewage can contain about 300 to 400 mg L^{-1} of organic compounds, 60% of which is readily degradable by bacteria commonly found in nature. Readily degradable implies that most of the material will be decomposed within about a week in a stream or other body of water that is sufficiently large. The change in the concentration of the organic matter with time is conveniently described by the first-order decay equation used to describe bacterial die-off. However, the value of this *K* depends on the specific organic compounds in the sewage. For domestic, sewage, an average value is about $K = 0.4$ per day, ranging from 0.1 to 0.7 per day. As more industrial wastes are contributed to the sewer system, the rate constant may increase or decrease. The amount of organic material discharged to a surface water depends on the population served by the municipal sewer system and treatment technology employed. Each person contributes about 90 grams per day of organics; thus, if the population is 100,000 people, the mass emission rate is 9 metric tons per day. The concentration of organics in the sewage depends on the amount of water added by the individual households and that added by other water uses in the community. If the average water use is 300 liters per person per day, the concentration is 300 mg L^{-1}.

Measurement of Potential Oxygen Demand of Organics in Sewage

COD

A parameter frequently used for industrial wastes, particularly where industrial wastes contribute heavily to the sewer system, is the *chemical oxygen demand* (COD), which is a measure of the amount of oxygen required to oxidize the organic matter—and possibly some inorganic materials—in a sample. Note that the method employed to obtain this parameter, which involves reflux of a sample in a strong acid with an excess of potassium dichromate, does not specifically measure the organic content in the sample, but rather the amount of oxygen required for oxidation. This approach therefore provides a direct measure of the potential impact of oxygen consumption on the oxygen content of the waterbody.

BOD

The *biochemical oxygen demand* (BOD) — is the most commonly used parameter in the analysis of oxygen resources in water. The BOD is the amount of oxygen consumed over time, usually 5 to 20 days, as the organic matter is oxidized both microbially and chemically.

DETERMINING BOD

The laboratory method used to determine BOD has changed very little since it was initiated in the 1930s. We begin by setting up many sample bottles to contain a sample of waste, mixed in water either from the disposal site or from a standard laboratory supply. Then we use standard methods to find the initial concentration of DO, after which the bottles are incubated in a dark water bath at a given temperature, usually 20°C. Every day for five or more days, we open a few of the bottles and measure the remaining DO. The difference between the initial value and the value at each time period, that is, the *demand*, is plotted as the BOD for the series of days.

From the data obtained in the laboratory, we construct a smooth curve that lets us calculate the reaction rate coefficient K by graphical or analytical methods. The curve we construct, whose equation is

$$BOD_t = BOD_L\ (1 - e^{-Kt}) \qquad ...(4)$$

becomes more an more horizontal as time progresses. By extrapolating the curve to horizontal, we can make an estimate of the ultimate value, called the *limiting value of BOD*, or *BOD_L*.

We can use the cure in this example to compute the value of K for the wastewater sample in this laboratory test. First, by extrapolating this curve to horizontal, we see that the estimated value of BOD_L would be about 7.8 mg L^{-1}. Next , by substituting the value read from the curve for day five (6.1 mg L^{-1}), that $BOD_t = BOD_5 = 6.1$ and $t = 5$. We an find K from Equation ...

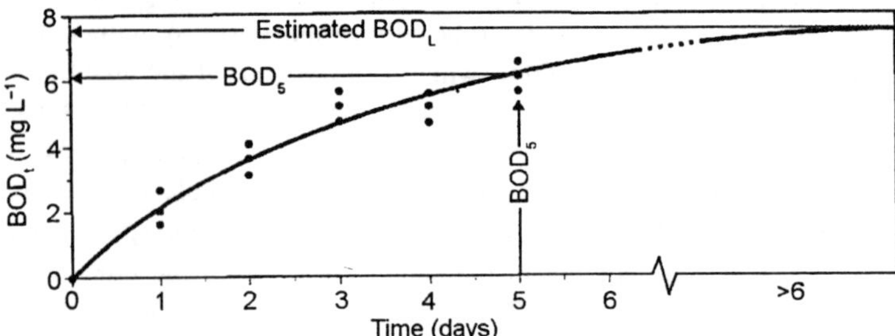

Fig. 5.6. Typical biochemical oxygen demand test results illustrating the graphical method for determining BOC_L and BOD_5 for use in solving eq. 4 to find K.

$$BOD_5 = BOD_L\,(1 - e^{-Kt})$$
$$6.1 = 7.6\,(1 - e^{-5K})$$

so that

$$1 - \frac{6.1}{7.6} = e^{-5K} \text{ or } \ln 0.197 = -5K$$

Thus

$$K = \frac{1.62}{-5} = 0.32 \text{ per day}$$

Impact of BOD on Dissolved Oxygen of Receiving Waters

Typically, municipal sewage is treated to some degree to remove organics, and hence to reduce BOD, before discharge to surface water waters in developed countries of the world. The amount of BOD remaining after treatment can range from 10 to 70% of the amount originally in the sewage. The impact on receiving waters depends on many environmental and waste characteristics, most of which are briefly mentioned later in this chapter. After the BOD parameters already explained, the next most important considerations are the amount of DO in the waterbody before the sewage is added, called the *initial DO* (DO_i), and the rate at which additional oxygen is transferred from the atmosphere to the receiving water.

Many rivers and streams have depressed DO concentrations, that is, a *DO deficit*, because of wastewater added by cities upstream. The saturation value of DO is 100% with the quantification in mg/L depending upon temperature, salinity, atmospheric pressure. The local deficit–the difference between the saturation value and the observed initial value at the location of waste discharge–must be included in the computation of the downstream Do deficit caused by a new effluent discharged to the stream. Before presenting an example calculation, let's first examine how nature deals with the DO deficit.

Deficits tend to be redressed by oxygen gas derived from the atmosphere. Such replenishment occurs by a process of gas-liquid mass transfer at the surface and subsequent mixing throughout the depth of water. This overall process can be described in an approximate manner by a first order-equation, the solution of which is

$$D_t = D_i e^{-RT} \quad \text{...(5)}$$

where

D_t is the deficit at time t

D_i is the initial deficit (i.e., DO_s $-DO_t$)

R is the reaeration coefficient

Note that the larger the magnitude of R, the quicker a given deficit is removed. The value of R depends on the degree of vertical mixing in a waterbody, as well as its overall depth. It varies form 0.1 per day in small ponds to above 1 in rapidly moving streams.

Most DO problem situations require that we determine the oxygen deficit resulting from the simultaneous effects of the oxygen demand of a waste and the competing restoration of oxygen from atmospheric reaeration. The combined result is termed the *self-purification capacity* of the waterbody. The deficit shown graphically as a function of time (or distance) is known as the *oxygen sag curve* because of its characteristic spoon shape. The equation for the curve is

$$D_t = \left[\frac{K(BOD_L)}{R-K}\right](e^{-Kt} - e^{-Rt}) + D_i e^{-} \qquad ...(6)$$

where the terms are as defined above.

The values of K and BOD_L are taken from figure; the value for D_i, the initial deficit, was 3 mg L^{-1} due to upstream discharges; and a reaeration coefficient R of 0.5 per day was chosen for a large, slow-flowing stream. This example might represent a case of highly treated sewage, a typical effluent plume after an initial dilution of 4, or poorly treated sewage dispersed from a diffuser that provides an initial dilution factor of 25.

One interesting result from the solution of the oxygen sag equation is that the deficit, irrespective of the saturation value of DO, is the same as long as the initial deficit is the same. Sometimes, water quality standards impose limits both on the amount of deficit *per se*

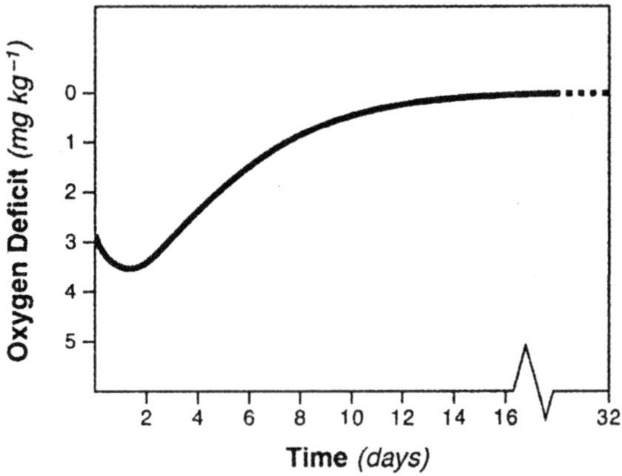

Fig. 5.7. Dissolved oxygen sag curve determined using values of BOD_L and K.

and on the resulting Do value itself. For example, a regulation could require that a waste discharge must neither increase the deficit by more than 10% nor depress resulting DO concentration below 5 mg L^{-1}.

When water systems are heavily used or highly valued water uses are threatened, additional factors require consideration:

1. The diurnal demand and supplies of oxygen from photosynthesizing organisms.
2. The oxygen demand of organic materials deposited in the sediment layer.
3. The oxygen demand of nitrogen compounds discharged in the effluent.
4. Variation in the reaeration coefficient R with travel time due to flow conditions in the waterbody.
5. The wide range of sewage flows encountered over the lifetime of a river.

Comprehensive computer programs are available to describe the concentration of oxygen in large watersheds consisting of dozens of interconnecting streams and dozens of wastewater inputs; however, data on plume travel, *K* values, and *R* values still have to be obtained with time-consuming laboratory studies and physically demanding field studies. Dilution of wastes is one of the most important factors to consider in assessing impacts. The methods used to assess effects on the dissolved oxygen resource of a waterbody and bacterial contamination are applicable to a wide variety of toxic chemical problems.

DILUTION OF EFFLUENTS

Dilution can be —but is not necessarily–an effective way to prevent pollution of surface waters. Environmental scientists do not categorically embrace the old saw, "Dilution is the solution to pollution." Instead, we recognize through the analytical process of risk assessment that certain principles underlie the utility of dilution in managing waste discharges. The first principle concerns the concentration dependence of the pollutant response mechanism. Here, we want to know if the effect of the contaminant is directly related to its concentration. That is, we ask if the concentration is reduced sufficiently, will the degree of effect be directly reduced?

Further, once the contaminant concentration is reduced, can it subsequently become more concentrated? Reconcentration is a phenomenon often associated with sediments and persistent organic chemicals such as PCBs. Dilution of waste streams containing high concentrations of suspended solids may prevent significant pollution near the discharge site, but such sediments may eventually settle out of the water column and concentrate in depressions in the streambed, where they can cause a variety of problems. Even in the ocean, waste disposal can lead to accumulation of sediments in the seabed. With persistent organic chemicals, whose solubility in water is low and whose affinity for sorption to animal tissues is high, adverse bioaccumulation can occur even from highly diluted mixtures. Thus, dilution may or may not solve a potential concentration-related problem.

Dilution in Streams and Rivers

Aside from concentration-toxicity considerations, social and economic considerations enter into the decision to use dilution. In addition, its use is also dependent on the availability of a sufficient quantity of dilution water. In the arid southwestern United States, for example, many streams are ephemeral; that is, they contain water only after major rain-fall events. In other streams in arid climates, the effluent form municipal treatment plants is the predominant flow for more than 50% of the year; that is, they are effluent-dominated streams. In such situations, dilution is somewhere between small and nil, except during storm runoff. For example, if the stream flow is 10 million liters per day and the effluent flow is 30 million liters per day, the contaminants dissolved in the effluent will be reduced in concentration by just 25%, assuming the concentration of each contaminant upstream is zero (generally, the effective dilution will be even less than that because there is almost always some measurable concentration of contaminants upstream).

The general equation used in determining the concentration after dilution is

$$c_f = \frac{c_e v_e + c_a v_a}{v_e + v_a} \qquad ...(7)$$

where

c_f = cross-sectional average final concentration in the stream

c_e = concentration in the effluent

v_e = volume flux of the effluent

c_a = concentration in the ambient dilution water upstream

v_a = volume flux of the ambient dilution water

In large rivers, the amount of dilution achieved depends on the method used to discharge the effluent into the river. To maximize the dilution, it is necessary to employ a diffuser, which consists of a pipeline with many exit orifices across the width of the river. Thus, if the river flow is 120 MLD, an effluent discharge of 30 MLD yields an 80% reduction in concentration of contaminants if the ambient concentration is zero. That is, the final concentration would be one-fifth the effluent concentration. Of course, the ambient concentration is almost always greater than zero, so Equation (7) must be used to estimate the final concentration accurately.

In many streams, construction of diffusers may be either inappropriate or prohibited. Consequently, amount of the dilution depends on natural mixing processes that occur during stream flow. But in large streams, the effluent plume may hug the bank for many miles, so it is not actively diluted with the main flow. In cases like this, it is not possible to estimate a range of values for the dilution rate. It is reasonable to assume that without a physical structure, like a diffuser, to mix the effluent into the river, we may consider only the natural die-off process in assessing the impact of bacterial contamination on downstream water uses. Similarly, if there is no dilution of the BOD, we can count only on the decomposition of organics and reaeration to restore or maintain the DO resource of the river.

Dilution in Large Bodies of Water

Large bodies of water, particularly open coastal waters, offer much greater opportunity for effective dilution of waste streams. Effective initial dilution is achieved by a multiport diffuser on the end of the outfall discharge pipe. Because the density of most wastewaters is very close to that of freshwater, the discharge of effluent to deep marine waters creates a strong buoyant force. Thus, the effluent, no matter how deep the discharge, will rise to the surface of the sea if it is not trapped by density gradients below the surface. As it rises toward the surface, the effluent effectively mixes with the surrounding ambient water, resulting in more and more dilution. A good analogy is the increasing width of a smoke plume as it rises in the atmosphere. The dilution is proportional to the square of the plume width.

In the effective placement of ocean outfalls, however, depth is not the only determining factor. All other things being equal, the greater the extent of vertical travel, the greater the amount of initial dilution. But those "other things" must be truly equal. If, for example, a location chosen for its great depth has poor circulation, the net result may be less effective dilution of wastes than that offered by placement in shallower, but more open, water. Such considerations are a major concern in the placement of outfalls in fjords, bays, and sometimes, estuaries.

Depth does not always provide the same opportunity for greater initial dilution in lakes and reservoirs because the difference in density between wastewater and receiving waters may be very small. In fact, it is not uncommon for industrial wastes to have a density greater than that of lake waters of these wastes tend to settle along the bottom rather than rise to the surface. However, because of their high temperatures, the cooling waters from large thermal-electric power stations have a density less than that of most lake waters. Thus, a deep discharge site can be advantageous for achieving effective reduction of thermal effects.

Many countries still allow the practice of dumping partially treated municipal sewage sludge into the ocean. For many years, before being banned in the United States in the 1980s, such sludge from New York City and Philadelphia was dumped in the Atlantic Ocean. Typically a portion of the sludge is particulate matter possessing a sufficiently high density to settle to the seabed, although currents and turbulence spread the material throughout the disposal zone. Other materials disposed of in the ocean, such as dredged sediments from harbors and waterways and some industrial wastes, behave similarly to sewage sludge.

Initial dilution and transport

The term *initial dilution* specifically identifies the amount of dilution achieved in a plume owing to the combined effects of the momentum and buoyancy-induced mixing of the fluid discharged from the orifice. This term is used both in regulatory practice and in plume hydrodynamics. The rate of dilution caused by these forces is quite rapid in the first few minutes after exiting the orifice, and then decreases markedly after the momentum and buoyancy are dissipated. Ambient currents also influence the rate of dilution during the buoyant rise of the plume irrespective of momentum and buoyancy. As current speed increases of so does initial dilution. In many cases, an initial dilution of 100 to 1, commonly sought in design of outfalls, is sufficient to reduce the toxicity of chemical contaminants to an acceptable level. (Note : when bacterial die-off is an important consideration, the distance from a designated use area is usually a more important factor than initial dilution per se. In this case, the time it takes microbes to travel a long distance increases the likelihood that they will be inactivated.)

Following initial dilution, waste streams undergo additional dilution, or dispersion, as they are transported by ambient currents and mixed with the surrounding water by turbulence. The process is analogous to the dispersion that takes place when smoke plumes dissipate in the atmosphere after the smoke has risen to an equilibrium level. In some aquatic systems, however, the effluent plume is not as easily observed.

Most modern coastal cities employ multiport ocean outfalls far offshore to protect beaches and nearshore recreational areas from the effects of bacterial contamination. These outfalls are frequently designed to maintain a diluted waste stream below the surface of the sea. Such systems are especially useful during the summer recreational season because they keep the immediate area of discharge free of unsightly messes. Moreover, they reduce landward transport of the diluted waste by onshore wind currents toward peak beach activities. The disadvantage of subsurface trapping lies in the fact that initial dilution is reduced compared with plumes rising to the surface, but this disadvantage is offset by the reduced risk of onshore transport.

Measurements and calculations

The dilution achieved in ambient transport of waste stream plumes in large water bodies can be described by physical laws—essentially the same laws used for describing aqueous flow in groundwater and gaseous flow in the atmosphere. However, we cannot solve the equations completely for the general case, because we lack existing data. That is, data obtained form field studies or from reports of previous studies are needed for empirical coefficients in the equations. However, many computer programs are available to obtain *approximate* solutions of the equations for complex cases involving multiple waste inputs and variable current speeds. In addition, we can sometimes use simplifications of the governing equations for many pollution assessment problems to obtain satisfactory estimates of contaminant concentration as a function of travel time or distance.

One simplified equation that has been used successfully over the past thirty years for large bodies of water gives us the maximum concentration at a distance *X*:

$$C_{\max} = C_{pi} erf \sqrt{\frac{Ub^2}{16\varepsilon_0 X}} \qquad \text{...(8)}$$

where:

$c_{\max}$ = centerline (maximum) concentration at distance *X*

c_{pi} = plume concentration at the end of initial dilution erf(#) = standard error function of (#)

U = current speed in the X direction

b = width in the Y direction (orthogonal to X) at the end of initial dilution

ε_0 = constant horizontal (Y direction) eddy diffusivity

X = travel distance [note that U/X can be replaced by $1/t$ time)]

In using this equation, it is important to use, values that are expressed in consistent units. For example the parameters U, b and ε_0 also contain a time unit. Another way to appreciate this requirement is to recognize that the argument *arg* of the error function must be dimensionless. The standard error function (erf) serves here as a mathematical representation of the way contaminants are observed to vary laterally (Y-direction) as the plume is transported in the X-direction. It describes the normal distribution curve used in evaluating variance around a mean value. Values can be e found from a tabular listing in a handbook or by using the standard error function in a spreadsheet program. The *transport dilution factor* is equal to the reciprocal of the value of erf(*arg*).

In a typical problem involving finding the transport dilution factor, we might estimate the highest concentration that would occur near the beach from an outfall 8 km (i.e., 8×10^5 cm) offshore when the onshore current speed U is 15 cm s^{-1}. The width b of the plume after initial dilution, which would be obtained from local observations, would likely be about 1,000 m (i.e., 10^5 cm). We can use a commonly cited value for the eddy diffusivity of $\varepsilon_0 = 10^4$ cm^2 s^{-1}. Now substituting these values into Equation 9) we get:

$$c_{\max} = c_{pi} erf\sqrt{1.17} = 0.87\ c_{pi} \qquad \text{...(9)}$$

The result is typically surprising: it shows that even with a travel distance of 8 km in a large, open body of water, the concentration is diluted to only 87%, resulting in a transport dilution factor of 1.2 (Contrast this factor of 1.2 to an initial dilution factor of about 100, which is what we expect for an ocean outfall).

This example shows us that initial dilution is often more important in reducing harmful levels than in dilution due to transport. This is true not only for BOD and DO problems, but also for most contaminants, including metals, ammonia, and toxic organics, that are found in partially treated effluents. However, in the case of indicator bacteria, the value of transport distance is realized, as demonstrated in the following example.

Assume an outfall whose degree of treatment is only 90% effective, so that instead of the 100,000 coliforms per 100 mL used in the example on the freshwater die-off of indicator organisms, the initial count of coliform is 100,000,000 per 100 mL. The initial dilution of 100:1 would reduce the concentration to 1% of the initial count, or 1,000,000. Then, using the reciprocal of the transport dilution factor found above (0.87), we can calculate that transport to the beach would further reduce the count to C_0 = 870,000 per 100 mL. The 8-km distance would be covered in t = 14.8 hours at a speed of 15 cm s^{-1}. If the die-off rate constant K is found to be above 0.46 per hour, near the usually expected lower range of values, the beach concentration of bacteria (C_t = 14.8), after substituting the values for C_0, K and t into Equation (3), would be about 960 per 100 mL.

Even with some form of secondary treatment, some regulatory agencies require disinfection to reduce bacterial concentration to bathing water standards. The result in the foregoing example demonstrates that chlorinated of the effluent might not be necessary with 90% removal, a level typically achieved in secondary treatment.

Because ocean outfalls often provide greater travel distances, coastal communities may not have to use chlorine disinfection in the sewage treatment plant to reduce microbial contamination of human use areas of the marine environment. This is important because use of chlorine can be hazardous. Also, chlorine may combine with organic materials in the sewage to produce compounds harmful to marine organisms and to the people who consume those organisms. The distance required for indicator organisms to die off to an acceptably low level depends on current speed and direction, as well as on the bacterial concentration at the end of the rapid initial dilution processes.

Dye Tracing of Plumes

Frequently, an oceanographic study is necessary to measure the bacterial concentration in the drifting plume as the current carries the water toward shore. Such studies usually employ a tracer, or dye, which is added to the sewage so that the plume can be followed for several kilometers. The data obtained can then be used to calculate a die-off rate constant, which may be useful for predicting the bacterial concentrations at different distances under a variety of current conditions.

Dye tracing is a well-known technique commonly used in hydraulic models and prototype field settings, although in deep outfall situations, tracers can be quite costly because of the large volumetric flow rates

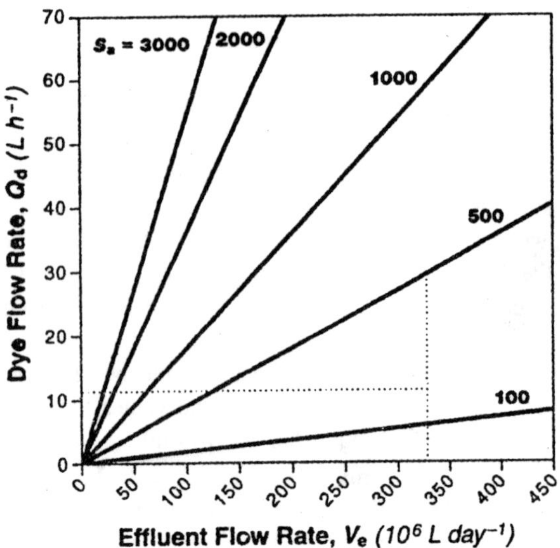

Fig. 5.8. A graph used to determine the dye required to provide 1µg L^{-1} in diluted effluent as used in the study.

and large dilutions usually achieved within a short time frame. The rate of dye addition Q_d to the effluent flow V_e needed to provide a dye concentration of C_d following dilution of S_a is

$$Q_d = \frac{V_e C_d \alpha_a S_a}{W \alpha_d} \qquad \text{...(10)}$$

where :

α_a = specific gravity of the diluted plume

α_d = specific gravity of the dye solution

W = weight fraction of dye in stock solution

Figure 5.8 shows the required dye rate in liters per hour for various dilution factors, and effluent flows in million liters per day, to achieve an ambient dye concentration of 1 µg L^{-1} in seawater. Rhodamine WT, typically used in dye studies is available as a 20% solution (∞_d = 1.19) in small (57-L) drums. Fluorometers used in field sampling can easily detect this dye at concentration of 0.5 to 1 µg L^{-1}.

To estimate the amount of dye needed to trace an effluent flow in a waterbody or any similar aquatic mixing question. If the flows marked on the x-axis of the graph and the slanted lines representing the dilution factors do not match exactly with the problems we have, Equation 10 may be used to refine the estimate.

Suppose, we have an effluent of 330 million liters (86.7 MGD) per day. A regulatory permit requires the effluent to be diluted by a factor of 200 at the end of a mixing zone. Suppose we set up our sampling boat at the mixing zone boundary and hope to measure 1 μg L^{-1} of dye as the plume passes under our boat. How much dye do we need to add to the effluent? Using the graph, estimate 330 along the abscissa and draw a lien up to the dilution factor line S_a = 500. Now estimate where S_a = 200 would lie on that line. From that point draw a line horizontally to the ordinate and estimate the dye requirement as 12 L h^{-1}.

For a more precise estimate, we will use Equation (10). First convert V_e = 330 million liters per day (*i.e.*, 330 × 10^6 L day^{-1}) to 13.7 × 10^6 liters per hour (L h^{-1}). For C_d, use 10^{-9} g dye per gram of seawater (this is approximately equal to 1 μg L^{-1}, assuming a specific gravity of 1). The specific gravity of sewage effluent diluted 100:1 with seawater is ∞_d = 1.023. (If this were a discharge to fresh water, the specific gravity would be 1.0). Using S_a = 200, W = 0.2, and ∞_d = 1.19 for rhodamine WT, as cited above, and substituting into Equation (10), we obtain Q_d = 12 L h^{-1}, verifying our estimate.

Spatial and Temporal Variation of Plume Concentrations

The concentrations of water quality indicators, such as bacteria, are neither uniform for steady with respect to the space and time scales involved in regulating the concentrations at the end of the mixing zone. In general, we assume that the concentrations of constituents in the horizontal extent of a plume from an outfall diffuser are uniform. But we can make no such assumption about the vertical direction. Vertical nonuniformity is commonly encountered in design, performance analysis and compliance monitoring, although in rivers it is not nearly the problem it is in estuaries, coastal water, and some lakes and reservoirs. Generally associated with density stratification in the receiving water, vertical nonuniformity is also associated with transport of a plume in a relatively thin lens as compared to the depth of the water column. For instance, if the plume is traveling on the surface, its constituents will be dispersed downward, and as these constituents disperse into the water column, the concentration of pollutants near the bottom edge of the plume gradually becomes less than that at the surface. (Thus if a permit condition requires that a maximum value be reported, sampling should be done at the surface, not at mid-depth.) Similarly, the dilution water mixed with the effluent being discharge is also vertically variable due to physical processes influencing the advection of ambient water into the region of the discharge. Dissolved oxygen (DO) is an example of one water-quality indicator that exhibits vertical nonuniformity in many riverine impoundments (reservoirs), lake, estuarine, and coastal situations.

Some transport and dispersion models produce estimates in terms of the *centerline concentration*, which is the maximum concentration for the cross section of the plume at a given distance downstream from the orifice. As the plume width expands with increasing distance, the maximum concentration progressively decreases. For example, the centerline (maximum) concentration at a distance of 60 meters from the diffuser may be 100 mg L^{-1}, while at 120 meters form the orifice, the maximum concentration would be closer to 70 mg L^{-1}. Other models calculate an average concentration for the cross section of the plume, and this of course also decreases downstream: the average concentration is always smaller than the maximum concentration. Both values need to be considered in field or lab verification studies, and both values may be useful for regulatory purposes.

Compliance Monitoring

Water pollution regulatory practice in the United States is founded on a system of discharge permits—known as the National Pollution Discharge Elimination System (NPDES) permits. Holders of these permits, e.g., municipal sewage treatment authorities and industries, must comply with the restrictions and requirements of their particular permit, such as limits on concentrations and mass emission rates of specific constituents. They also have to meet water quality standards established for the waterbody into which they discharge their effluents. Some permits, especially for coastal water discharges, require elaborate environmental monitoring projects. The permit holder is required to conduct monitoring activities and report the results to demonstrate compliance with permit conditions. On occasion, regulatory agencies conduct studies to verify and revise ongoing programs. Monitoring data reflect the wide variations of conditions found in the natural environment, and discharges and regulators are often challenged to rationalize monitoring results with predictions used in setting permit conditions.

Mixing Zones

Permit conditions of regulatory agencies usually allow exceptions to one or more of the water quality criteria within a mixing zone adjacent to the point of discharge. A *mixing zone* might be established by purely arbitrary considerations or by use of data and simulations with mathematical models. Many mixing zone determinations are made on the basis of the expected dilution rate that will be provided by efficient diffuser designs intended to optimize initial dilution. For large bodies of water, a common approach is to describe the width of the zone as the depth of water at the disposal site and the length as the length of the diffuser. For large rivers, it is common to restrict a mixing zone so it does not extend completely across the river, thereby leaving a "safe passage" that lets aquatic species avoid high concentration of wastewater constituents. But sometimes the shape of a mixing zone is entirely arbitrary, say, a rectangular zone downstream from a discharge pipe equal to one-fourth the width of the stream and extending downstream for one kilometer. Frequently, there are two or three mixing zones for different groups of contaminants and degrees of toxicity.

Regulatory Use

Regulatory interest may be appropriately directly toward both discrete and average values of contaminants. For example the state of California and the U.S. EPA specify maximum allowable instantaneous values for some parameters as well as several temporal average values (e.g., 30-day and 6-month arithmetic means). In some cases, these regulations are based on knowledge of the effects on aquatic organisms. In other cases, these values are specified to acquire statistics on the performance of the wastewater treatment plant.

Criteria that are expressed in terms of temporal averages (daily to semiannual) suggest that plume concentrations be assessed extensively in three dimensions, both at the boundary of the mixing one and, in some cases, at sensitive biological resource locations down current. Current speed and direction play significant roles when assessing the concentrations at the boundary. By incorporating data on the cyclical variation of effluent composition, density profiles, and current direction, it is possible to construct a running 6-month average (or median) for a number of points on the mixing zone boundary. The 6-month average is expected to be quite variable at these points, and the point with the highest exposure frequency may not have the highest average concentration.

Beyond the mixing zone, there may be regions where current streams of diluted effluent, each leaving the zone at a different time in different direction, would converge over a reef, a kelp forest, or a swimming area. In this case, the frequency and duration of exposure may be more important than the highest observed concentration in assessing the overall impact on these resources.

Verification Sampling

Aside from the question of whether discrete values or cross-sectional averages are used to test compliance with criteria, the way in which field samples are used to verify or compare with model results is an important considerations.

In laboratory or field verification studies of plume performance, the average value is measured or captured in a sample bottle only by chance. Characteristically, the field value measured is form a very small spatial region and represents a signal over a certain time span. Many samples are sought form the same cross section in order to arithmetically compute an average. In the laboratory, using hydraulic models, this is relatively easy to do. But in the field, where multiple plumes are usually involved, sampling is more complicated. We're usually trying to take samples from a moving flow field too deep below the surface to see, using a moving sampler mounted on a moving boat. It is therefore reasonable to assume some uncertainty as to what portion of the cross section the value represents.

For these reasons, field verification studies of submerged plumes in deep rivers, lakes, and coastal waters are best attempted for a cross section as far from the outfall as practical, as long as the region is still within the range where the plume is continuous. Nearer to the outfall, the values are changing more readily and the dimensions of the plume are much smaller, making it much harder to get the sampler in the right place or even in the plume. In addition, it is best to conduct the study when currents are low, so that the plume rises nearest to the surface. Placement of the sampling device may be improved because it may even possible to see the plume. Aside from the ease of sampling, samples taken during low currents may be especially useful for verification of regulatory compliance.

Subsurface Pollution

Freshwater comprises approximately 3% of all water on Earth. Approximately 95% of this small fraction occurs as water in the subsurface, i.e., groundwater. This, groundwater is a critical resource throughout the world. Groundwater is a major source of potable water, supports food and crops production, and used for myriad industrial activities. As such, the availability, quality, and sustainability of groundwater resources are issues of great significance.

Groundwater as a Resource

Groundwater has long been used by humankind. This likely occurred initially through the use of natural springs and later via hand-dug wells. There are numerous references to groundwater in ancien texts, such as the works of Plato and Aristole. Chinese archaeologists have found wells in the Hunan Province dating back to 2000 BC. Evidence of hand-dug wells has been found at archaeological sites thousands of years old.

Ground water use has increased greatly in the past several decades. Groundwater use worldwide has tripled in the past 50 years. This is also true for the U.S. These increases are a direct result of increases in population and economic development. A primary use of groundwater is to supply potable water for drinking and other domestic use. Groundwater serves as a significant source of potable water throughout the world, ranging form 15% in Australia to 75% in Europe. In the United States, groundwater provides approximately half of the total potable water supply. The percentage of the U.S. population relying on groundwater as their primary source of potable water varies greatly by state.

In addition to supplying potable water, groundwater is also used for many other purposes. Agricultural applications, primarily as irrigation for crops, constitute the single largest use of groundwater in the U.S. The majority of irrigation use occurs in the semi-arid and arid regions of the U.S., as would expect. Others uses of groundwater include industrial, mining, and power generation.

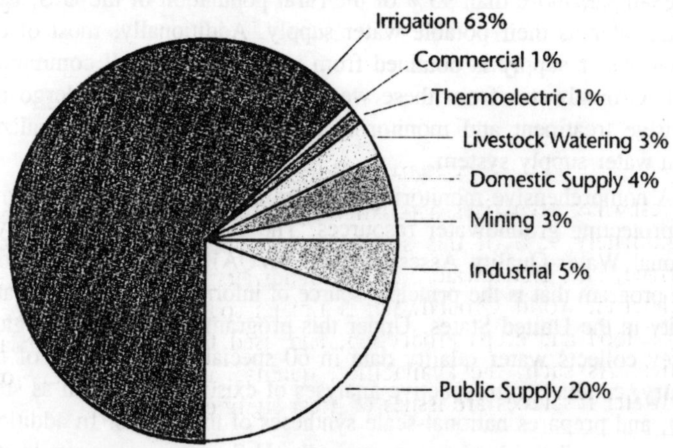

Fig. 5.9. National ground water use.

Groundwater Pollution

Hand in hand with the use of groundwater by humans is the pollution of groundwater by humans. This pollution can occur in many forms, including hazardous industrial organic compounds, fuel components, heavy metals, agrochemicals, pathogenic microorganisms, and salinity. There are two general categories of groundwater contamination: those produced from point source and those that develop form diffuse or nonpoint sources.

The growth of population centers with the attendant increase in population densities and industrial/ commercial development, had an enormous impact on groundwater use and quality. Water-resource development during the evolution of an urban center follows a typical pattern. Generally, increasing demand for potable water that occurs as population increases results in over-pumping of groundwater from the original well field located in the city center. The amount of water being extracted is greater than the amount recharged, causing a decline in water levels. This eventually requires the city to supplement their potable water supply from other sources. An ancillary effect of the over-pumping and falling water levels is subsidence of the land surface. Another significant effect of over-pumping in coastal areas in seawater intrusion.

The development of high-intensity agriculture and the widespread use of fertilizes and pesticides have lead to major groundwater pollution issues for rural areas. This is of particular concern because groundwater generally is the predominant source of potable water for rural areas. For example, more than 95% of the rural population of the U.S. uses groundwater as their potable water supply. Additionally, most of the potable water supply is obtained from individual or small community wells. Groundwater from these wells does not typically undergo the extensive treatment and monitoring that is prevalent for centralized urban water supply system.

A comprehensive monitoring program is instrumental in managing and protecting groundwater resources. The U.S. Geological Survey's National Water-Quality Assessment (NAWQA) Program is a nationwide program that is the principal source of information on groundwater quality in the United States. Under this program, the U.S. Geological Survey collects water quality data in 60 special study regions of the country, conducts retrospective analyses of existing data (such as state data), and prepares national-scale syntheses of the results. In addition, many state and local governments in the U.S. carry out groundwater monitoring programs. Specialized monitoring programs are conducted in association with characterization and remediation of contaminated sites.

Groundwater Pollution Risk Assessment

Physical properties of the subsurface result in significant differences in the behavior of groundwater compared to that of surface water. For example, residence times for groundwater range from years to hundreds of years or more; these times are much grater than those for streams. Dilutions effects, either in water or the atmosphere, are much less significant for groundwater compared to surface-water systems. In addition, the absence of light eliminates are possibility of photochemical reactions, which are a major route of transformation for above ground systems. The nets result is that once groundwater and the subsurface are contaminated, it is very difficult to decontaminate. Thus, pollution prevention is critical to maintain sustainable groundwater resources. Groundwater pollution risk assessment is a key aspect of pollution prevention.

Preventing contamination from entering the environment is the only sure way of preventing pollution. Laws and regulations promulgated during the past few decades have helped to reduce the overall contaminant load to the environment. Obviously, however, preventing all contamination from entering the environment is not possible, hence the need for groundwater pollution risk assessment, the goal of which is to evaluate the risk posed by a given activity or event to groundwater resources. Implementing risk assessments enhances the effective management of groundwater resources and helps to minimize potential contamination.

Groundwater pollution risk is composed of two components: groundwater vulnerability and contaminant load. Groundwater vulnerability is the intrinsic susceptibility of the specific aquifer in question to contamination. Several factors affect groundwater vulnerability. An aquifer that is close to ground surface, overlain by sandy soil, and located in an area with high precipitation rates would clearly be more vulnerable to contamination than an aquifer that is hundreds of meters below ground surface and located in an area of low precipitation.

Factors involved in the contaminant load are the type of contaminant, the amount of contaminant released, the time scale of release, and the mode of release. The pollution potential of a contaminant is controlled by its transport and fate behavior.

Contaminants that are transported readily (e.g., those with high aqueous solubility and low sorption) and that are not transformed to any great extent (i.e., are persistent) generally have greater potential to pollute groundwater. For most contamination events, the contamination enters the environment in close proximity to land surface (e.g surface spills, leaking storage tanks, and landfills). Thus, transport of contaminants from the source zone to groundwater necessitates trave through the soil and vadose zone. Attenuation processes such as sorption and biodegradation can act to reduce and limit the transport of contaminants to groundwater. For this reason, the soil and vadose zone is often refereed to as a "*living filter.*" The degree to which contaminants will be attenuated is a function of the type of contaminant and the nature of the subsurface. Generally, the greater the amount of contamination released, the greater the pollution potential. The timescale and mode of release from buried storage tanks may be more prone to cause groundwater contamination than release from tanks stored above ground on concrete pads.

The risk of groundwater pollution results form the contamination of intrinsic vulnerability and contaminant load factors. Thus, an aquifer that is very vulnerable may have little to no risk of pollution if the contaminant load remains negligible. Conversely, an aquifer that has a relatively low degree of vulnerability may have a significant pollution risk if the contaminant load factor is very high. The greatest pollution risk will be associated with locations where the aquifer has a high vulnerability and the contaminant loading is high.

In general, not much can be done to modify or change the inherent vulnerability of an aquifer. That is why it is critical to focus on controlling the contaminant load for managing and preventing groundwater pollution. This can be done through implementing land-use and facility-operation regulations. For example, aquifer vulnerability assessments can be used in the siting of new facilities that involve production, storage, or disposal or hazardous materials. Including aquifer vulnerability as a siting factor can prevent building such facilities in locations where groundwater is most susceptible to pollution. For existing facilities, operation procedures can be implemented to minimize the production and disposal of wastes.

Special procedures and tools have been developed to assess groundwater pollution risk. This is usually done by employing a geographic information system. The first step typically involves constructing an aquifer vulnerability map. This map incorporates one or more factors that influence aquifer vulnerability. A commonly used aquifer vulnerability tool is DRASTIC. The map was developed using soil drainage as the aquifer vulnerability factor, and fraction of cropland acreage, population density, and nitrogen loading as the loading (land use) factors.

Point-Source Contamination

Point-source systems are characterized by very localized contamination release. Primary examples include surface spills, leaking storage tanks, disposal pits, and waste-injection wells. The distribution of contamination at sites associated with point-source release follows a general pattern. The region of the subsurface where the majority of the original contamination is present in referred to as the *source zone*. It is usually in close proximity to the location of the contaminant release. The source zone generally encompasses a relatively small area and contains the majority of the contaminant. Conversely, the region where groundwater is contaminated by dissolved compounds originating from the source zone is referred to as a *groundwater contaminant plume*. The plume is often much larger than the source zone. However, the amount of contaminant mass associated with the plume may represent a small fraction compared to the mass in the source zone.

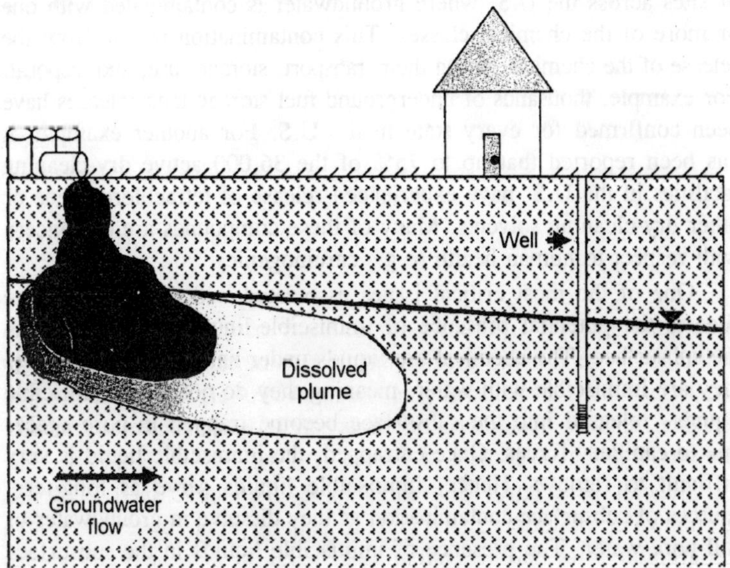

Fig. 5.10. Distribution of contamination at a point-source groundwater contamination site.

The configuration or "architecture" of the source zone (e.g., porous-medium heterogeneity, total contaminant mass, contaminant distribution) and source-zone "dynamics" (e.g., mass-transfer processes, transformation processes) is central to the pollution risk posed by the site. For example, the magnitude (size and concentration level) of the groundwater contaminant plume generated from the source zone is clearly dependent on the magnitude and rate of contaminant mass transfer from the source zone to surrounding regions (i.e. the source-zone mass flux). This mass transfer will be influenced by groundwater flow patterns (which are mediated by the physical properties of the porous media), and by the type, amount, and distribution of contaminant. The groundwater contaminant plume is generally the primary source of human-health risk posed by these types of contaminated sites, as use of contaminated groundwater is usually the major route of potential exposure to subsurface contamination. Thus, it is critical to characterize source-zone and contaminant plume properties at a given site.

Hazardous Organic Chemicals

The contamination of groundwater by hazardous organic chemicals and the associated risk to human health and the environmental is one of the primary groundwater pollution issues facing the U.S. and other industrialized countries. It is estimated that there are tens of thousands of sites across the U.S. where groundwater is contaminated with one or more of the chemical classes. This contamination results from the release of the chemical during their transport, storage, use, and disposal. For example, thousands of underground fuel storage tank releases have been confirmed for every state in the U.S. For another example, it has been reported that up to 75% of the 36,000 active dry-cleaning facilities in the U.S. have experienced release of tetrachloroethene or other dry-cleaning solvents. Many of these compounds are of concern with respect to human health (e.g., carcinogenic).

One of the most critical issues associated with hazardous waste sites is the potential presence of immiscible-liquid contamination in the subsurface. The chemical are liquids under natural conditions, and they are immiscible with water, meaning they do not mix. Immiscible liquids released into the subsurface become trapped in pore spaces due to capillary forces. Once entrapped, the immiscible liquid is very difficult to remove. Hence, immiscible liquids serve as long-term sources of subsurface contamination as they dissolve in groundwater or soil-pore water. The presence of immiscible liquids in the subsurface of a site can greatly impact the costs and time required for site remediation. For example, for sites contaminated by dense nonaqueous phase liquids (DNAPLs), it is estimated that upwards the hundreds of years may be necessary to achieve health-based groundwater cleanup objectives using standard pump-and-treat systems (ITRC). This clearly illustrates the critical importance of addressing immiscible-liquid contamination when it is present at a site. Unfortunately, as is widely acknowledged, cleaning up sites contaminated by immiscible liquids is one of the greatest challenges in the field of environmental remediation. In fact, according to several reviews conducted by expert panels convened by the National Research Council, the presence of immiscible liquids is usually the single most important factor limiting the cleanup of organic-chemicals contaminated sites. In addition, the presence of immiscible liquids greatly complicates site characterization risk assessment efforts. The estimated cost to clean up the immiscible-liquid contaminated sites in the U.S. is $100 billion or more.

The distribution of immiscible liquids in the subsurface is controlled by the physical and chemical properties of the porous media and by the properties of the chemical. A primary property of concern is the density of the immiscible liquid in comparison to that of water. In fact, this is such an important property that the immiscible liquids are classified based on whether they are more (DNAPL) or less (LNAPL) dense than water, where NAPL = nonaqueous phase liquid. Because DNAPLs are denser than water, they can sink below the water table with sufficient volume of release. A conceptual diagram of the distribution of a DNAPL in the subsurface is presented. In contrast, to DNAPLS, LNAPLS float on the water table because they are less dense than water. The distribution and amount of immiscible liquid present in the source zone has a major impact on the nature of the groundwater contaminant plume that forms at the site.

In addition to source-zone properties, the size of the dissolved-contaminant plume will be influenced by the nature of the chemicals and by physical, chemical, and biological properties of the subsurface. Plumes are generally relatively small for chemicals that undergo significant attenuation. Conversely, very large contaminant plumes can form for chemicals that undergo minimal attenuation. For example, groundwater plumes comprised of chlorinated-solvent constituents are typically hundreds to thousands of meters long because these compounds are difficult to biodegradable and have relatively low sorption. Conversely, plumes generated from fuel hydrocarbons such as gasoline are generally much shooter, primarily because the compounds are more amenable to biodegradation.

Landfills

The contamination of soil and groundwater from landfills is another important problem that continues to threaten groundwater resources, posing risks to human health and the environment. Chemicals, both hazardous and nonhazardous can be leached from the materials that are disposed of in landfills. This landfill-derived contamination of *leachate* can enter the soil and migrate through the vadose zone, eventually contaminating groundwater resources. Landfills are used for the disposal of numerous types of wastes, garbage, and material. Thus, contamination emanating form landfills can contain mixture of numerous types of compounds, increasing the complexity of the pollution problem.

Contaminated water or leachate is produced when water (e.g., precipitation or irrigation) enters the landfills and contacts the waste materials. The infiltrating water can remove hazardous and nonhazardous chemicals, including metals, minerals, slats, organic chemicals (e.g., chlorinated solvents, petroleum hydrocarbons, and pesticides), and various other toxic compounds. Millions of gallons of leachate can percolate through a landfill, depending on the size of the landfill or disposal facility. A well-designed landfill should contain the disposed materials, isolating potential harmful leachate form the environment and specifically from groundwater and drinking water resources. However, it is impossible to contain and control all wastes produced at landfill and disposal sites. Leachate can enter groundwater systems as a result of poorly designed or improperly constructed landfills, deterioration of landfill liners, and landfills constructed without liners (e.g. typically older designs).

Given that leachate is likely to leak to some degree form all landfills, it is essential to implement a well-designed monitoring scheme to help manage landfill pollution problems. For example, a series of monitoring wells can be constructed around the perimeter of the landfill. The presence of high salt concentrations (e.g., Cl^-) indicates that the potential threat of contamination to groundwater. Similarly, total dissolved solids (TDS) can be used to indicate potential contamination of groundwater by landfill leachate. Salts serve as good indicators of leachate contamination because of their high mobility and persistence, which means they usually constitute the leading (downgradient) front of the groundwater leachate plume. Monitoring pH is another means to detect potential contamination of groundwater by leachate. Generally, the pH of landfill leachate is lower than that of uncontaminated groundwater. A third method to detect potential leachate contamination of groundwater is by determining the oxidation-reduction potential. Highly reducing conditions typically indicate either low pH or high microbial degradation activity. Waste materials disposed off in landfills are often subject to microbially mediated decay and decomposition. The microorganisms consume oxygen during the degradation process, thus reducing the oxidation-reduction potential. In addition, some microorganisms can degrade waste under anaerobic conditions and in the process release methane and other gases. Thus, the presence of such gases in groundwater is another indicator of landfill-leachate contamination.

As noted above, landfill leachate comprises many compounds. Thus, the transport and fate behavior of leachate is complex and highly variable. Compounds with high mobilities and low degradation potentials (high persistence) will tend to be transported much further than compounds that sorb to soil and/or that are easily degraded. For this reason, it is difficult to predict the transport and fate behavior of leachate in soil and groundwater.

Diffuse-Source Contamination

Contamination resulting form nonpoint (diffuse) source of pollution refers to those inputs that occur over a wide area and are associated with particular land sues. This is in contrast to point-source discharges, which occur form a specific, very localized source such as a leaking fuel tank or pipe. Nonpoint sources generally encompass much larger scales (regional and even global scales) and as a result can create very large zones of pollution compared to point sources. However, the contaminant concentrations associated with nonpoint source pollution are generally lower than those associated with point sources. This section will focus on two major diffuse-source issues: agrochemical contamination and salt-water intrusion.

Agrochemical Pollution of the Subsurface

The advent of intensive agricultural practices during the last century and into the 21st century has greatly increased global food production. However, it has also greatly increased the use of fertilizers and pesticides, so called *agrochemicals*. The increasing use of these agrochemicals has lead to extensive pollution of groundwater. For example, nitrate, derived from fertilizers, pesticides, and animal wastes, is one of the most widespread and pervasive groundwater contaminants in the United States and the world. In a survey of almost 200,00 water sampling reports, the EPA found that more than 2 million people were using water from public potable-water supply system for which nitrate standards were exceeded at least once between 1986 and 1995. An additional 3.8 million people were using water form private wells that exceed federal drinking water standards. Researchers predict, due to past and current inputs of nitrates into the environment, that the full effect of overapplication of nitrate fertilizers will not be realized for another 30-40 years.

As we might expect, agricultural areas generally have the most significant groundwater nitrate contamination problems. For example, major agricultural regions such as the San Joaquin Valley in California and the Ogallala aquifer system extending form Minnesota to Texas are areas with high vulnerability to nitrate groundwater contamination. In addition, these agricultural regions are also experiencing severe declines in groundwater levels as a result of excessive groundwater extraction for irrigation. It is common in agricultural areas to see the compound effect of declining groundwater levels coinciding with high levels of nitrate contamination.

Nitrate is generally very mobile in the subsurface. This, in conjunction with the large areal extent of input, results in extensive groundwater plumes of nitrate contamination. It is difficult and expensive to clean up groundwater once it is contaminated by nitrate. For example, the costs associated with managing nitrate contamination problems in California and Iowa are estimated to exceed $200 million per year. A standard approach for dealing with shallow groundwater nitrate contamination is to drill deeper wells. However, this can be done only a limited number of times. Another common practice is to blend contamination water with uncontaminated water. The objective of this technique is to dilute the nitrate to concentrations below the drinking water standard. This approach increases the overall use of water resources. It is estimated that closing down a well due to nitrate contamination, and drilling another well or blending contaminated water with cleaner supplies, can cost between $200,000-500,000 per well. Point-of-use treatment for nitrate is also expensive, requiring methods such as reverse osmosis. A key to solving the problem of nitrate groundwater contamination is to prevent future contamination by using best management practices.

Another major class of agrochemicals, also widely used in agricultural practices, are pesticides. Pesticides are typically applied at the land surface, usually as a chemical spray. Once applied to the ground surface, pesticides can migrate downward through the vadose zone with infiltrating water and contaminate groundwater. Pesticides are used throughout the world, primarily for agriculture, to control weeds, insects, and fungal pests. A study conducted between 1991 and 2001 by the U.S. Geological survey found that 42% of wells sampled in agricultural regions across the United States contained the common pesticides atrazine and diethylatrazine. Furthermore, it was reported that about 20% of the wells sampled in major aquifer systems throughout the U.S. contained both atrazine and diethylatrazine. These results illustrate the extent of groundwater pollution by pesticides.

The majority of pesticides are organic compounds. Their transport and fate behavior in the subsurface is a function of their chemical properties. Some pesticides are relatively mobile (e.g., 2,4-D), while others are highly sorbed (e.g., DDT). The classes of pesticides that were initially developed, such as DDT, are very persistent in the environment. Newer pesticides have been designed in part to be less persistent. A complicating factor in the evaluation of pesticide pollution problems is the sheer number of pesticides available for use. It is not customary to analyze for all possible pesticide compounds, their derivatives, and possible degradation products in groundwater monitoring surveys. In addition, analytical limitations have constrained detection capacity. However, recent advancement have resulted in more frequent detection of pesticide compounds in groundwater supplies.

Salt-Water Intrusion

Salinization of freshwater is one of the most serious and widespread groundwater contamination issues throughout the world. Areas along coasts where seas or oceans meet continental landmasses and island systems are most vulnerable to salinization of groundwater resources. In most coastal settings, groundwater beneath and surface generally exists of a lens of "freshwater." The freshwater is separated from the denser seawater by a diffuse interface known as the freshwater/saltwater interface or zone of dispersion. The freshwater/saltwater interface typically resides at some point near where the ocean meets the land mass (coastline) and extends vertically in the subsurface, separating freshwater (inland) from high salinity water (seawater). This freshwater lens tends to become thinner as it approaches the shoreline. If recharge into the aquifer equals extraction rates due to pumping, the freshwater/ saltwater interface will remain stable. However, if groundwater pumping exceeds recharge, the saline water will invade the freshwater aquifer and the freshwater/saltwater interface will progress further inland. As this zone progresses further inland, groundwater supply wells can become contaminated from the invading saltwater. This phenomenon is known as *saltwater intrusion*, and it has caused severe degradation and contamination of groundwater.

It does not take very much saltwater to contaminate a freshwater groundwater supply. Only 3-4% addition of salinity can make a fresh water supply unsuitable for most uses, including drinking water and even irrigation. As little as 6% addition of saltwater will render a freshwater source (groundwater) unsuitable for any use except cooling or flushing purposes. Once a freshwater resource is degraded by salt contamination, it will take a very long time for that aquifer to recover, and if positive groundwater recharge conditions are not re-established, it may never recover. Remediation efforts are often cost-prohibitive if not impossible due to the technical constraints associated with removing or decreasing the levels of salt concentration in groundwater. The first response is often abandonment of the contaminated wells, accompanied by drinking of new wells further inland. Effectively, the freshwater resource will have been lost, and supplying new water depends on availability of groundwater supplies further inland. Furthermore, if excessive groundwater pumping continues, the saltwater will continue to invade further inland, again contaminating fresh groundwater supplies. The other option currently available is to construct a desalinization plant to treat the contaminated groundwater prior to use. Such plants are likely to increase in use in the future, as technology improves and the availability of water supplies dwindles.

Although coastal regions may be most susceptible to saltwater intrusion, many noncoastal areas are also being affected by salinization of groundwater. Many natural geological systems can lead to salinization of freshwater resources. Areas once occupied by deep-water oceans or seas are now part of continents. These areas, now deep within the subsurface, contain ancient geologic units that contain high concentrations of salt or brine water. These salt-containing geologic units can contaminate freshwater aquifers when over pumping occurs in the region, causing the water table to decline and encroach into the high-salinity geologic unit.

Generally there are no harmful health effects associated with low concentration of chloride in drinking water. In some cases, salt (chloride) can be harmful to people with heart or kidney conditions. The EPA has set unenforceable secondary drinking water guidelines for chloride at 250 mg/L. However, the contamination of water by saltwater intrusion will increase salt concentrations far beyond what can be tolerated by humans. The primary concern with the contamination of drinking water supplies from saltwater intrusion is the large-scale loss of water resources.

Other Groundwater Contamination Problems

Although we have discussed some of the major groups of contaminants threatening groundwater resources, such as hazardous organic chemicals (e.g. chlorianted solvents and fuel-type hydrocarbons), agrochemical pollutants (e.g., nitrates and pesticides), and salinization (e.g., saltwater intrusion and high-salinity groundwater), it is important to note some other contaminants that also present potential threats to the quality of groundwater supplies.

Pathogen Contamination of Groundwater

Contamination of groundwater by microbial pathogens, including viruses, bacteria, and protozoa, is of significant concern throughout the world.

Potential sources of pathogens for groundwater contamination include land disposal of sewage treatment by-products (wastewater, biosolids), septic tank systems, and latrines. Risks posed by pathogen-contaminated groundwater are generally believed not to be significant for public supply system, given the level of treatment applied before use. Of much greater concern is potential pathogen contamination of groundwater used for private water supplies, because water from private wells typically undergoes little or not treatment before use. Thus, residential areas with septic systems and private wells are particularly susceptible to potential effect of groundwater contamination by pathogens.

Proper siting and construction of septic and well systems is necessary to minimize potential pollution problems. So called wellhead protection rules have been developed to prevent the siting or application of pathogen sources too close to water-supply wells. Several recent surveys of groundwater across the U.S. have shown that the incidence of human viruses in groundwater is greater than previously believed and may in part be due to septic tank system.

Gasoline Additive: Methyl Tertiary-Butyl Ether (MTBE)

In the mid 1990 it was discovered that methyl tertiary-butyl-ether (MTBE), additive in gasoline, had caused extensive contamination of groundwater throughout the United States. MTBE had been used since the early 1970s as an oxygenate to promote more efficient combustion of fuel in automobiles. Ironically, while the intended use of MTBE has in fact reduced toxic emissions (e.g., carbon monoxide) released to the atmosphere from automobiles, it has contributed to the widespread contamination of groundwater resources from leaking underground storage tanks, posing serious threats to the quality of drinking water supplies. The contamination of groundwater by MTBE is extensive in the U.S. The EPA reports, citing a study of Chevron, that MTBE concentrations exceeded 100 $\mu g\ L^{-1}$ in 47% of 251 California sites surveyed, 63% of 153 Texas sites surveyed, and 81% of 41 Maryland sites surveyed. Responding to the rising occurrences of MTBE contamination, California in 1999 became the first state to ban the use of MTBE in gasoline reformulation after 2002. The EPA has now begun regulatory action to phase out the use of MTBE in gasoline.

The primary concern associated with the release of MTBE to the subsurface environment (from spills or leaking underground fuel storage tanks) it its mobility and persistence. For example, at most fuel station sites affected by leaking storage tanks, it is commonly observed that MTBE has migrated much farther than typical gasoline contaminants (e.g., benzene, toluene). Unlike these contaminants, MTBE has a relatively low biodegradation potential. The health effects associated with MTBE through human ingestion of drinking water are not well understood nor well documented. Due to lack of appropriate data on the health effects of MTBE contamination, the EPA has been unable to define national regulatory standards based on quantitative estimates for health risks. However, the EPA has set provisional drinking water health advisory limits at 20–40 $\mu g\ L^{-1}$.

Solvent Additives: 1,4-Dioxane

In recent years, 1,4-dioxane (dioxane) has gained considerable attention as a primary "*emerging*" contaminant for groundwater resources. Dioxane, like MTBE, is commonly used as an additive. It is a synthetic organic chemical used as an industrial solvent or solvent "stabilizer" that prevents the breakdown of chlorinated solvents during manufacturing processes. Dioxane is commonly added to chlorinated solvents and other solvents such as tetrachlorethene (PCE), trichloroethene (TCE), 1,1,1-trichlorethane (TCA), and paint thinners. Dioxane is also used as a solvent for the manufacturing of paper, cotton, textiles, and various organic products, automotive coolant, shampoos, and cosmetics. It is estimated that TCE and TCA contain approximately 1% and 2-8%, 1,4-dioxane, respectively. It was estimated that between 10 and 18 million pounds of 1,4-dioxane were produced in the U.S. in 1990.

Although 1,4-dioxane has been used as a stabilizer for solvents since the 1940s, the extensive contamination of groundwater by dioxane was not documented until the mid-1990, when improved analytical methods allowed for the detection of lower concentrations. Dioxane is a very mobile and persistent compound, and is listed as a probable human carcinogen. The EPA has not yet defined a national regulatory standard for dioxane. Action levels have been adopted by some states for 1,4-dioxane. For example, the California Department of Health Services has set an action level of 3 $\mu g\ L^{-1}$.

Perchlorate in Groundwater

Perchlorate, another "emerging" groundwater contaminant, was first detected in drinking water in 1997 and has since been recognized to pose a significant threat to groundwater resources. As mentioned previously, it often requires a significant advancement in analytical capability to first observe an emerging contaminant's presence in the environment and a in particular groundwater. The development of an analytical method to detect low concentrations of perchlorate allowed for the recognition of its widespread occurrence in groundwater in the U.S. Perchlorate is an inorganic anion and is often present as a salt complex or ammonium salt as ammonium perchlorate. Perchlorate is used for numerous industrial and military purposes. For example, it is a primary constituent tin the manufacturing and use of rocket propellants (solid rocket fuel) and other explosives. For this reason, much of the perchlorate contamination of groundwater is derived from military bases and installations. In fact, it is estimated that approximately 90% of perchlorate compounds are produced for use in defense activities and the aerospace industry.

Perchlorate, like many salt compounds, is extremely mobile in groundwater. In addition, perchlorate is not readily susceptible to chemical or microbial degradation and is thus extremely persistent in the environment. In studies perchlorate has been shown to interfere with the uptake of iodine by hormone regulation, metabolism regulation, fetus development, and child development. To date, the EPA has not established a maximum contaminant level of enforceable regulatory limit for perchlorate in drinking water. A few states have adopted "action levels" or advisory levels for perchlorate, which generally range from 1 $\mu g\ L^{-1}$ to 6 $\mu g\ L^{-1}$.

Arsenic in Groundwater

Contamination of groundwater by arsenic (As) is another important groundwater contaminant problem throughout the world. The high toxicity associated with arsenic is of primary concern for human health.

As a result of the high associated toxicity of arsenic, regulatory standards have recently been lowered by the EPA, from 50 $\mu g\ L^{-1}$ to 10 $\mu g\ L^{-1}$. The additional water-treatment costs associated with meeting the revised standard are projected to be in the billions of dollars.

The contamination of groundwater and drinking water can result form natural or human activities. Arsenic is a naturally occurring metallic element that is found in soil, rocks, air, plants, and animals. Arsenic in soil and rocks can act as sources for groundwater contamination. Through processes such as dissolution, weathering, and erosion, arsenic can be released into the environment, resulting in the contamination of groundwater and drinking water supplies. Arsenic sources associated with human activities include agriculture, use as a wood preservative, the burning of fuels and wastes, smelting and mining, paper production, glass manufacturing, and cement manufacturing. In 1997, almost 8 million pounds of arsenic were released to the environment by human activities. Extensive adverse health impacts due to arsenic contamination of groundwater have been recently documented in Bangladesh.

Acid-Mine Drainage

Another source of pollution that poses threat to groundwater is acid mine drainage. Highly acidic water is produced when rain or groundwater comes into contact with mine tailings or mining wastes. The highly acidic water can change the oxidation-reduction potential in the groundwater and may cause the release of minerals and metals, serving as a source of groundwater contamination. The geology, hydrology, and mining technology employed will determine the nature of the acid mine drainage.

Sustainability of Groundwater Resources

Clearly, the world's population is greatly dependant on groundwater for many uses. Thus, managing and protecting our groundwater resources is essential to ensuring that sufficient quantities of quality groundwater will be available for future generations. This concept is referred to as *maintaining long-term sustainability of groundwater resources*.

Ensuring groundwater sustainability requires balancing supply and demand. Hydrologic and geologic factors (climate, topography, subsurface properties) exert primary control on the intrinsic supply of groundwater. The potential impact of global climate change on the hydrologic cycle and groundwater supply is of concern. Groundwater pollution affects the fraction of the intrinsic supply that is of sufficient quality for use. Thus, it imposes a constraint on supply. The demand for groundwater is associated with land use and population density. The significant increase in groundwater use observed over the past few decades is a result of population growth and economic expansion.

The primary means by which to ensure long-term sustainability of groundwater resources is to mange supply and demand. The groundwater supply can be extended through moderating demand and can be supplemented with additional sources of water. Water reuse is one primary method being implemented to enhance sustainability. This includes reusing municipal wastewater directly, either as potable water or for secondary uses such as irrigation, or indirectly (e.g., artificial groundwater recharge). Instituting conservation measures to reduce demand is another method.

The use of external supplies is another means of supplementation. For example, the *central Arizona Project* (CAP) provides surface water from the Colorado River to supplement groundwater resources in Arizona. The CAP canal extends 336 miles form Lake Havasu City to Tucson, and cost $3.6 billion to construct. In Tucson, all water provided by the CAP is recharged into groundwater prior to being pumped to the surface and treated for potable use.

A potential problem with using external water supplies, versus water reuse and conservation, to manage supply and demand, is that it imposes a demand on water resources at the point of origin. An example of this is, in fact, the Colorado River, which barely exists as a river close to its entry into Mexico because of its great degree of use upstream. Thus, the demand has just been shifted in part form one location to another. Changes in land use also exert an impact on supply and demand. For example, groundwater use is now shifting from agriculture to residential as urban centres increase in size and population density.

Balancing supply and demand is encapsulated in the concept of *safe yield*. In essence, the principle of safe yield is that the amount of groundwater extracted should not exceed the amount replenished through recharge. This concept is simple in theory. Unfortunately, it is very complex in practice. A primary reason for this is that water resource issues are influenced to a great extent by nonscience factors (political, economic, societal, legal). This cause the management of demand among the competing uses prevalent in large urban centers to be a complex and difficult task.

In addition, planning for future water resource use is difficult and fraught with uncertainty. For example, there is always uncertainty in estimating the future supply of groundwater available in a region. This uncertainty is compounded by uncertainty in demand, which must be estimated based on future population growth and land-use patterns. These difficulties are exacerbated by the "invisibility" of groundwater–as opposed to surface water, it cannot be seen. The prevalent attitude of "out of sight, out of mind" often creates impediments to increasing the awareness of groundwater resources issues.

6

Water Treatment

The treatment of water may be divided into three major categories:

1. Purification for domestic use.
2. Treatment for specialized industrial applications
3. Treatment of wastewater to make it acceptable for release or reuse.

The type and degree of treatment are strongly dependent upon the source and intended use of the water. Water for domestic consumption must be thoroughly disinfected to eliminate disease-causing microorganisms, but may contain appreciable levels of dissolved calcium and magnesium (hardness). Water to be used in boilers may contain bacteria, but must be quite soft to prevent scale formation. Wastewater being discharged into a large river may require less rigorous treatment than water to be reused in an arid region. As world demand for limited water resources, grows, more sophisticated and extensive means will have to be employed to treat water.

Most physical and chemical processes used to treat water involve similar phenomena, regardless of their application to the three main categories of water treatment listed above. Therefore, after introductions to water treatment for municipal use, industrial use, and disposal, each major kind of treatment process is discussed as it applies to all of these applications.

Municipal Water Treatment

The function of a modern water treatment plants is to produce clear, safe, even tasteful drinking water from raw water that may consist of a murky liquid pumped from a polluted river laden with mud and swarming with bacteria; well water, much too hard for

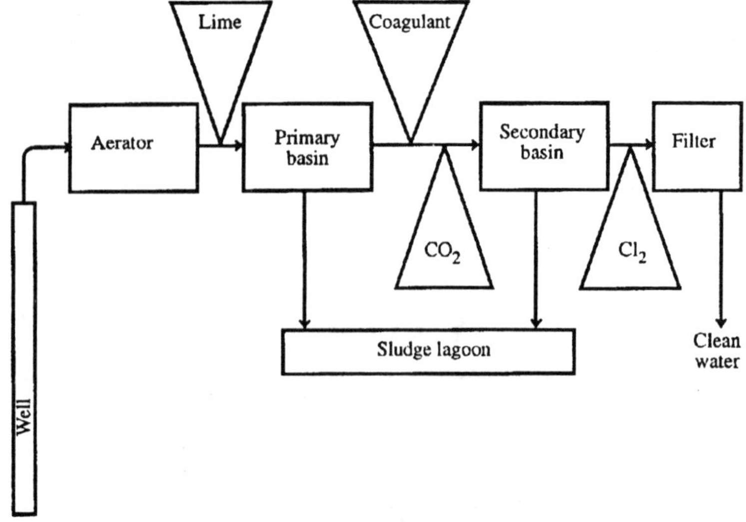

Fig. 6.1. Schematic of a municipal water treatment plant.

domestic use and containing high levels of stain-producing dissolved iron and manganese; or other questionable sources of raw water. This particular facility treats water containing excessive hardness and a high level of iron. The raw water taken form wells first goes to a narrator. Contact of the water with air removes volatile solutes such as hydrogen sulfide, carbon dioxide, methane, and volatile odorous substances such as thiomethane (CH_3SH) and bacterial metabolites. Contact with oxygen also aids iron removal by oxidizing soluble iron(II) to insoluble iron(III). After aeration, lime is added as CaO or $Ca(OH)_2$ to raise the pH, causing the formation of precipitates containing Ca^{2+} and Mg^{2+} hardness ions. These precipitates settle from the water in a primary basin. Much of the solid material remains in suspension and may be caused to settle by the addition of coagulants, such as iorn(III) and aluminum sulfates, which form gelatinous metal hydroxides, activated silica, or synthetic a organic polyelectrolytes. The setting of the coagulated solids occurs in a secondary basin after the addition of carbon dioxide to lower the pH. Sludge from both the primary and secondary basins is pumped to a sludge lagoon. The water is finally chlorinated, filtered, and pumped to the city water mains.

Treatment of Water for Industrial Use

Water is widely used in various process applications in industry. Other major industrial applications are boiler feedwater and cooling water. The kind and degree of treatment of water in these applications depends up on the end use. As examples, cooling water may require only minimal treatment, removal of corrosive substances and scale-forming solutes is essential for boiler feedwater and water used in food processing must before of pathogens and toxic substances. Improper treatment of industrial water can cause problems, such as corrosion, scale formation, reduced heat transfer in heat exchangers, reduced water flow, and product contamination. These effects may cause reduced equipment performance or equipment failure increased energy costs due to inefficient heat utilization or cooling, increased costs for pumping water, and product deterioration. Obviously, the effective treatment of water at minimum cost for industrial applications is a very important area of water treatment.

Numerous factors must betaken into consideration in designing and operating an industrial water treatment facility. These include the following:

1. Water requirement
2. Quantity and quality of available water sources
3. Sequential use of water (successive uses for applications requiring progressively lower water quality)
4. Water recycle
5. Discharge standards.

External treatment, usually applied to the plant's entire water supply, involves processes such as aeration, filtration, and clarification to remove material from water that may cause problems. Such substances include suspended or dissolved solids, hardness, and dissolved gases. Following this basic treatment, the water may be divided into different streams, some to be used without further treatment and the rest to be treated for specific applications.

Internal treatment is designed to modify the properties of water for specific applications. Examples of internal treatment include several processes, the most common of which are the following:

1. Removal of dissolved oxygen by reaction with hydrazine (N_2H_4) or sulfite (SO_3^{2-}):

$$2SO_3^{2-} + O_2 \rightarrow 2SO_4^{2-} \quad ...(1)$$

$$N_2H_4 + O_2 \rightarrow 2H_2O + N_2(g) \quad ...(2)$$

2. Addition of chelating agents to react with dissolved Ca^{2+} and prevent formation of calcium deposits.
3. Addition of precipitants, such as phosphate used for calcium removal
4. Treatment with dispersants to inhibit scale

5. Addition of corrosion inhibitors
6. Adjustment of pH
7. Disinfection for food processing uses or to prevent bacterial growth in cooling water.

Sewage Treatment

Typical municipal sewage contains oxygen-demanding materials, sediments, grease, oil, scum, pathogenic bacteria, viruses, salts, algal nutrients, pesticides, refractory organic compounds, heavy metals, and an astonishing variety of flotsam ranging form children's socks to sponges. It is the job of the waste treatment plant to remove as much oft his material as possible.

Characteristics used to describe sewage include turbidity (international turbidity units), suspended solids (ppm), total dissolved solids (ppm), acidity (H^+ ion concentration or pH), and dissolved oxygen (in ppm O_2). Biochemical oxygen demand is used as a measure of oxygen-demanding substances. Current processes for the treatment of wastewater may be divided into three main categories: primary treatment, secondary treatment, and tertiary treatment, each of which is discussed separately. Also discussed are total wastewater treatment systems, based largely upon physical and chemical processes.

Waste from a municipal water system is normally treated in a *publicly owned treatment works*, POTW. In the United States these systems are allowed to discharge only effluents that have attained a certain level of treatment, as mandated by federal law. As discussed later, one of the objectives of hazardous waste treatment is to remove water from the waste and purify it such that it can be safely discharged to a POTW.

Primary Waste Treatment

Primary treatment of wastewater consists of the removal of insoluble mater such as grit, grease, and scum from water. The first step in primary treatment is normally screening. Screening removes or reduces the size of trash and large solids that get into the sewage system. These solids are collected on screens and scraped off for subsequent disposal. Most screens are cleaned with power rakes. Comminuting devices shred and grind solids in the sewage. Particle size may be reduced to the extent that the particles can be returned to the sewage flow.

Grit in wastewater consists of such materials as sand and coffee grounds, which do not biodegrade well and generally have a high settling velocity. *Grit removal* is practiced to prevent its accumulation in other parts of the treatment system, to reduce clogging of pipes and other parts, and to protect moving parts from abrasion and wear. Grit normally is allowed to settle in a tank under conditions of low flow velocity, and it is then scraped mechanically from the bottom of the tank.

Primary sedimentation removes both settleable and floatable solids. During primary sedimentation it is important for flocculent particles to aggregate for better settling, a process that may be aided by the addition of chemicals. The material that floats in the primary settling basin, known collectively as grease, consists of oils, waxes, free fatty acids, and insoluble soaps containing calcium and magnesium. Normally, some of the grease settles with the sludge and some floats to the surface, where it may be removed by a skimming device.

Secondary Waste Treatment by Biological Processes

The most obvious harmful effect of biodegradable organic matter in wastewater in BOD, consisting of a biochemical oxygen demand for dissolved oxygen by microorganism-mediated degradation of the organic matter. *Secondary wastewater treatment* is designed to remove BOD, usually by taking advantage of the same kind of biological processes that would otherwise consume oxygen in water receiving the wastewater. Secondary treatment by biological processes takes many forms but consists basically of the following: Microorganisms provided with added oxygen are allowed to degrade organic material in solution or in suspension until the BOD of the waste has been reduced to acceptable levels. The waste is oxidized biologically under conditions controlled for optimum bacterial growth and at a site where this growth does not influence the environment.

A common biological process for treating wastewater consists of *fixed-film biological* (FFB) reactors. The oldest and simplest of these is the *trickling filter* in which wastewater is sprayed over rocks or other solid support material covered with microorganisms, such that

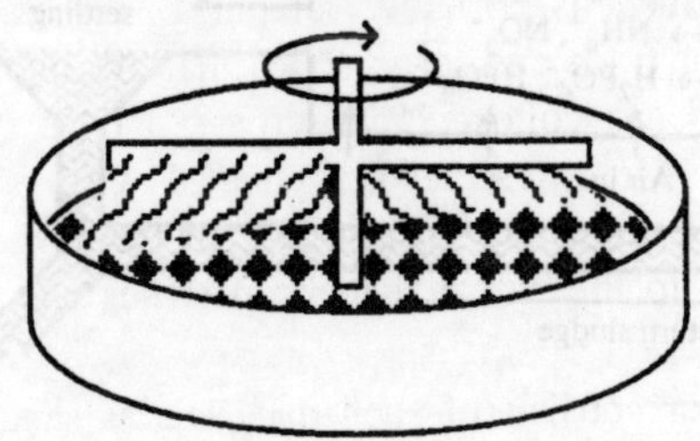

Fig. 6.2. Trickling filter for secondary waste treatment.

contact of the wastewater with air is allowed and degradation of organic matter occurs by the action of the microorganisms. A ore modern approach is the *rotating biological reactor*, consisting of groups of large plastic disks mounted close together on a rotating shaft such that growths of microorganisms on the disks are alternately exposed to air and wastewater. The greatest advantage of these processes is their low energy consumption because air does not have to be pumped into the system.

The *activated sludge process*, is probably the most versatile and effective of all waste treatment processes. Microorganisms is the aeration tank convert organic material in wastewater to microbial biomass and CO_2. Organic nitrogen is converted to ammonium ion or nitrate. Organic phosphorus is converted to orthophosphate. The microbial cell matter formed as part of the waste degradation processes is normally kept in the aeration tank until the microorganisms are past the log phase of growth at which point the cells flocculate relatively well to form settleable solids. These solids collect in the bottom part of a settler and a fraction of them is discarded. Part of the solids, the return sludge, is recycled to the head of the aeration tank and comes into contact with fresh sewage. The combination of a high concentration of "hungry" cells in the return sludge and a rich food source in the influent sewage provides optimum conditions for the rapid degradation of organic matter. The activated sludge process removes organic carbon from water by conversion to CO_2 and by incorporation into biomass.

The disposal of waste sludge from an activated sludge plant can be a problem, primarily because it is only about 1% solids and contains many undesirable components. Normally, partial water removal is accomplished by drying on sand filters, vacuum filtration, or

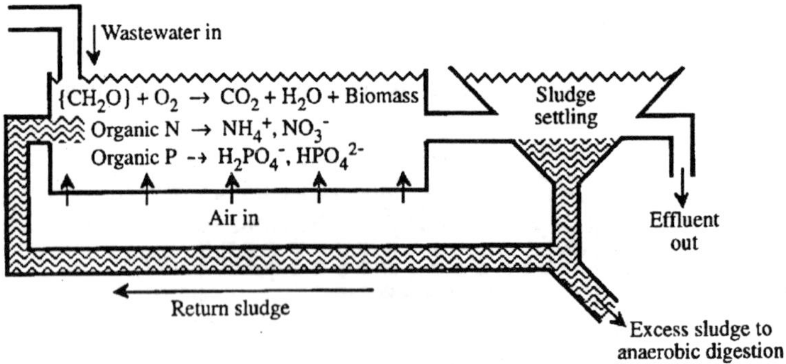

Fig. 6.3. Activated sludge process.

centrifugation. The dewatered sludge may be incinerated or used as landfill, both of which are facing increased regulatory and public opposition. To a certain extent, sewage sludge may be digested in the absence of oxygen by methane-producing anaerobic bacteria to produce methane and carbon dioxide,

$$2\{CH_2O\} \rightarrow CH_4 + CO_2 \quad ...(3)$$

a process that reduces both the volatile-matter content and the volume of the sludge by about 60%. A carefully designed plant may produce enough methane to provide for all of its power needs, which can be a very important factor in reducing costs.

One of the most desirable means of sludge disposal is to use it to fertilize and condition soil. However, care has to be taken that excessive levels of heavy metals are not applied to the soil as sludge contaminants.

Tertiary Waste Treatment

Tertiary waste treatment (sometimes called *advanced waste treatment*) is a term used to describe a variety of processes performed on the effluent from secondary waste treatment so that the water can be reused. The contaminants removed by tertiary waste treatment fall into the general categories of (1) suspended solids; (2) dissolved organic compounds; and (3) dissolved inorganic materials, including the important class of algal nutrients. Each of these categories presents its own problems with regard to water quality. Suspended solids are primarily responsible for residual biological oxygen demand in secondary sewage effluent waters. The dissolved organics are the most hazardous from the standpoint of potential toxicity. The major problem with dissolved inorganic materials is that presented by algal nutrients, primarily nitrates and phosphates. In addition, potentially hazardous toxic metals may be found along the dissolved inorganics.

In addition to these chemical contaminants, secondary sewage effluent often contains a number of disease-causing microorganisms, requiring disinfection in cases where humans may later come into contract with the water. Among the bacteria that may be found in secondary sewage effluent are organisms causing tuberculosis, dysenteric bacteria (*Bacillus dysenteriae*, *Shigella dysenteriae*, *Shigella paradysenteriae*, *Proteus vulgaris*), cholera bacteria (*Vibrio cholerae*), bacteria causing mud fever (*Leptospira icterohemorrhagiae*), and bacteria causing typhoid fever (*Salmonella typhosa*, *Salmonella paratyphi*). In addition, viruses causing diarrhea, eye infections, infectious hepatitis, and polio may be encountered. Ingestion of sewage still cause disease, even in more developed nations.

Physical-Chemical Treatment of Municipal Wastewater

Complete physical-chemical wastewater treatment systems, which rely upon chemicals and energy-intensive processes for water treatment, can be used instead of biological treatment systems. Around 1970 it appeared that such systems might become seriously competitive with biological waste treatment system. However, with the first "energy crisis" of 1973 the costs for energy and chemicals, many of which are based upon petroleum or remote energy-intensive processes for their manufacturing, made physical-chemical treatment of municipal wastewater much less attractive. Basically, a physical-chemical treatment process involves:

1. Removal of scum and solid objects.
2. Clarification, generally with the addition of a coagulant, and frequently with the addition of other chemicals (such as lime for phosphorus removal).
3. Filtration to remove filterable solids.
4. Activated carbon adsorption.
5. Disinfection.

Industrial Wastewater Treatment

Wastewater to be treated must be characterized fully, particularly with a thorough chemical analysis of possible waste constituents and their chemical and metabolic products. The biodegradability of wastewater constituents should also be determined.

One of two major ways of removing organic wastes is biological treatment by an activated sludge, or related process. It may be necessary to acclimate microorganisms to the degradation of constituents that are not normally biodegradable. Consideration needs to be given to possible hazards of biotreatment sludges, such as those containing excessive levels of heavy metal ions. The other major process for the removal of organics from wastewater is sorption by activated carbon, usually in columns of granular activated carbon. Activated carbon and biological treatment can be combined with the use of powdered activated carbon in the activated sludge process. The powdered activated carbon sorbs some constituents that may be toxic to microorganisms and is collected with the sludge. An important consideration in using activated carbon to treat wastewater is the hazard that spent activated carbon may present from the wastes it retains. These hazards may include those of toxicity or reactivity; for example, spent activated carbon used to sorb explosives manufacture wastes may be very dangerously reactive, especially after the carbon has dried. Regeneration of the carbon is expensive and can be hazardous in some cases.

Wastewater can be treated by a variety of chemical processes, including acid/base neutralization, precipitation, and oxidation/reduction. In some cases these treatment steps must precede biological treatment; for example, wastewater exhibiting extremes of pH must be neutralized in order for microorganisms to thrive in it. Cyanide in the wastewater may be oxidized with chlorine and organics with ozone, hydrogen peroxide promoted with ultraviolet radiation, or dissolved oxygen at high temperatures and pressures. Heavy metals may be precipitated with base, carbonate, or sulfide.

Wastewater can be treated by several physical processes. In some cases, simple density separation and sedimentation can be used to remove water-immiscible liquids and solids. Filtration is frequently required, and flotation by gas bubbles generated on particle surfaces may be useful. Wastewater solutes can be concentrated by evaporation, distillation, and membrane processes, including reverse osmosis, hyperfiltration, and ultrafiltration. Organic constituents can be removed by solvent extraction, air stripping, or steam stripping. Synthetic resins that attract organic constituents are useful for removing some pollutant solutes from wastewater. Cation exchange resins are effective for the removal of heavy metals.

Removal of Solids

Relatively large solid particles are removed from water by simple *settling* and *filtration*. A special type of filtration procedure known as *microstraining* is especially effective in the removal of the very small particles. These filters are woven from stainless steel wire so fine that it is barely visible. This enables preparation of filters with openings only 60-70 μm across. These openings may be reduced to 5-15 μm by partial clogging with small particles, such as bacterial cells. The cost of this treatment is likely to be substantially lower than the costs of competing processes. High flow rates at low back pressures are normally achieved.

The settling and filtration of colloidal solids from water usually require *coagulation*. Salts of aluminum and iron are the coagulants most often used in water treatment. Of these, alum or filter alum is most commonly used. This substances is a hydrated aluminum sulfate, $Al_2(SO_4)_3{\bullet}18H_2O$. When added to water, the aluminum ion in this salt hydrolyzes by reactions that consume alkalinity in the water, such as:

$$Al_2(H_2O)_6^{3+} + 3HCO_3^- \rightarrow Al(OH)_3(s) + 3CO_2 + 6H_2O \quad ...(3)$$

The gelatinous hydroxide thus formed carries suspended material with it as it settles. In addition, however, it is likely that positively charged hydroxyl-bridged dimers such as,

$$(H_2O)_4Al\begin{matrix}\overset{H}{O}\\ \underset{H}{O}\end{matrix}Al(H_2O)_4^{4+}$$

and higher polymers are formed that interact specifically with colloidal particles, bringing about coagulation. Metal ions in coagulants also react with virus proteins and destroy up to 99% of the virus in water. Iron salts that precipitate gelatinous $Fe(OH)_3(s)$ may be used as coagulants in a manner analogous to aluminum. Sodium silicate partially neutralized by acid aids coagulation, particularly when used with alum.

Natural and synthetic polyelectrolytes act to flocculate colloidal-size particles. Among the natural compounds so used are starch and cellulose derivatives, proteinaceous materials, and gums composed of polysaccharides. Synthetic polymers that are effective flocculants are now widely used. Neutral polymers and both anionic and cationic polyelectrolytes have been employed successfully as flocculants in various applications.

An important class of solids that must be removed from wastewater consists of suspended solids in secondary sewage effluent that arise primarily from sludge that was not removed in the settling process. These solids account for a large part of the BOD in the effluent and may interfere with other aspects of tertiary waste treatment. For example, these solids may clog membranes in reverse osmosis water treatment processes. The quantity of material involved may be rather high. Processes designed to remove suspended solids often will remove 10-20 mg/L of organic material from secondary sewage effluent. In addition, a small amount of the inorganic material is removed as well.

Removal of Calcium and other Metals

Calcium and magnesium salts, which generally are present in water as bicarbonates or sulfates, cause water hardness manifested by the formation of an insoluble "curd" from the reaction of soap with Ca^{2+} and Mg^{2+}. Another problem caused by hard water is the formation of mineral deposits. For example, when water containing calcium and bicarbonate ions is heated, insoluble calcium carbonate is formed:

$$Ca^{2+} + 2HCO_3^- \rightarrow CaCO_3(s) + CO_2(g) + H_2O \quad ...(4)$$

This product coats the surfaces of hot-water systems, clogging pipes and reducing heating efficiency. Dissolved salts such as calcium and magnesium bicarbonates and sulfates can be especially damaging in boiler feedwater. Clearly, the removal of water hardness—called *water softening*—is essential for many uses of water:

Several processes are used for softening water. On a large scale in a centralized water treatment plant the lime-soda process utilizing lime, $Ca(OH)_2$, and soda ash, Na_2CO_3, is employed. Calcium is precipitated as $CaCO_3$ and magnesium as $Mg(OH)_2$. When the calcium is present primarily as "bicarbonate hardness," it can be removed by the addition of $Ca(OH)_2$ alone:

$$Ca^{2+} + 2HCO_3^- + Ca(OH)_2 \rightarrow 2CaCO_3(s) + 2H_2O \quad ...(5)$$

When bicarbonate ion is not present at substantial levels, a source of CO_3^{2-} must be provided at a relatively high pH by the addition of Na_2CO_3. For example, calcium present as the chloride can be removed from water by the addition of soda ash:

$$Ca^{2+} + 2Cl^- + 2Na^+ + CO_3^{2-} \rightarrow CaCO_3(s) + 2Cl^- + 2Na^+ \quad ...(6)$$

The precipitation of magnesium as the hydroxide requires a higher pH than the precipitation of calcium as the carbonate:

$$Mg^{2+} + 2OH^- \rightarrow Mg(OH)_2(s) \quad ...(7)$$

The high pH required may be provided by the basic carbonate ion from soda ash:

$$CO_3^{2-} + H_2O \rightarrow HCO_3^- + OH \quad ...(8)$$

The water softened by lime-soda softening plants usually suffers from two defects. First, because of super-saturation effects, some $CaCO_3$ and $Mg(OH)_2$ usually remain in solution. If not removed, these compounds will precipitate at a later time and cause harmful deposits or undesirable cloudiness in water. The second problem results from the use of highly basic sodium carbonate, which gives the water is *recarbonated* by bubbling CO_2 into it. The carbon dioxide neutralizes excess OH^- to bring the pH within the range of 7.5-8.5 and converts the slightly soluble calcium carbonate and magnesium hydroxide to their soluble bicarbonate forms:

$$CaCO_3(s) + CO_2 + H_2O \rightarrow Ca^{2+} + 2HCO_3^- \quad ...(9)$$

$$Mg(OH)_2(s) + 2CO_2 \rightarrow Mg^{2+} + 2HCO_3^- \quad ...(10)$$

Water adjusted to a pH, alkalinity, and Ca^{2+} concentration very close to $CaCO_3$ saturation is *chemically stabilized*. It neither deposits $CaCO_3$ is water pipes, which can clog them, nor dissolves protective $CaCO_3$ coatings from the pipe surfaces. With a Ca^{2+} concentration much below $CaCO_3$ saturation, water is called *aggressive*.

Calcium may be removed from water very efficiently by the addition of orthophosphate:

$$5Ca^{2+} + 3PO_4^{3-} + OH^- \rightarrow Ca_5OH(PO_4)_3(s) \qquad ...(11)$$

It should be pointed out that the chemical formation of a slightly soluble product for the removal of undesired solutes such as hardness ions, phosphate, iron, and manganese must be followed by sedimentation in a suitable apparatus. Frequently, coagulants must be added, and filtration employed for complete removal of these sediments.

Water may be softened by ion exchange, the reversible transfer of ions between aquatic solution and a solid material capable of bonding ions. The water is passed over a solid cation exchanger in the sodium ion form, represented by $Na^{+-}\{Cat(s)\}$, which trades the Na^+ ions bound with the solid for Ca^{2+} ions in water.

$$2Na^{+-}\{Cat(s)\} + Ca^{2+} \rightarrow Ca^{2+-}\{Ca^{2+-}\{Cat(s)\}_2 + 2Na^+ \qquad ...(12)$$

The Na^+ ions released cause no problems with the end use of the water, such as for laundry.

Water softening by cation exchange is widely used, effective, and economical, although it does cause some deterioration of water quality arising from the contamination of wastewater by sodium chloride. Such contamination results fro the periodic need to regenerate a water softener with sodium chloride, in order to displace calcium and magnesium ions form the resin and replace these hardness ions with sodium ions:

$$Ca^{2+-}\{Cat(s)\}_2 + 2Na^+ + 2Cl^- \rightarrow 2Na^{+-}\{Cat(s)\} + Ca^{2+} + 2Cl^-$$

During the regeneration process, a large excess of sodium chloride must be used—several kilograms for a home water softener, Appreciable amounts of dissolved sodium chloride can be introduced into sewage by this route.

Chelation, or as it sometimes known, *sequestration*, is an effective method of softening water without actually having to remove calcium and magnesium from solution. A complexing agent is added that binds with and greatly reduces the concentrations of free hydrated cations, as shown by the chelation of calcium ion with excess EDTA anion (Y^{4-}):

$$Ca^{2+} + Y^{4-} \rightarrow CaY^{2-} \qquad ...(14)$$

This reduces the concentration of hydrated calcium ion, preventing the precipitation of calcium carbonate:

$$Ca^{2+} + CO_3^{2-} \rightarrow CaCO_3(s) \qquad ...(15)$$

Removal of Iron, Manganese, and Heavy Metals

Soluble iron and, to a lesser extent, manganese are found in many groundwaters because of reducing conditions that favor the soluble +2 oxidation state of these metals. Both metals are detrimental to water quality because of their staining tendencies. The basic method for removing both of these metals depends upon oxidation to higher insoluble oxidation states. The oxidation is generally accomplished by aeration and is favored by a high pH.

Heavy metals such as copper, cadmium, mercury, and lead are found in wastewaters from a number of industrial processes. Because of the toxicity of many heavy metals, their concentrations must be reduced to very low levels prior to release of the wastewater. A number of approaches are used in heavy metals removal.

Lime treatment for water softening discussed above removes heavy metals as insoluble hydroxides, basic salts, or coprecipitates along with calcium carbonate or ferric hydroxide. This process does not completely remove mercury, cadmium, or lead, so their removal is aided by the addition of sulfide, taking advantage of the tendency of heavy metals to act as sulfide-seekers:

$$Cd^{2+} + S^{2-} \rightarrow CdS(s) \qquad \text{...(16)}$$

Lime precipitation does not normally permit recovery of metals, produces large quantities of potentially hazardous sludge, and is sometimes undesirable from the economic viewpoint.

Electrodeposition (reduction of metal ions to metal by electrons at an electrode), *reverse osmosis*, and *ion exchange* are frequently used for metal removal. Solvent extraction using organic-soluble chelating substances is also effective in removing many metals. *Cementation*, a process by which a metal deposits by reaction of its ion with a more readily oxidized metal, may be employed:

$$Cu^{2+} + Fe\ (\text{iron scrap}) \rightarrow Fe^{2+} + Cu \qquad \text{...(17)}$$

Activated carbon adsorption effectively removes some metals from water at the part per million level. Sometimes a chelating agent is sorbed to the charcoal to increase metal removal.

Even when not specifically designed for the removal of heavy metals, most waste treatment processes remove appreciable quantities of the more troublesome heavy metals encountered in wastewater. These metals accumulate in the sludge from biological treatment, so sludge disposal must be given careful consideration.

Removal of Dissolved Organics

Very low levels of exotic organic compounds in drinking water are suspected of contributing to cancer and other maladies. Some of these re chlorinated organic compounds produced by chlorination of organics in water, especially humic substances. Removal of organics to very low levels prior to chlorination has been found to be effective in preventing formation of compounds of the chloroform ($CHCl_3$) type called trihalomethanes. In addition, many organic compounds survive, or are produced by, secondary wastewater treatment.

The standard method for the removal of dissolved organic material is adsorption on granular or powdered activated carbon, a material that is made from a variety of carbonaceous materials, including wood, pulp-mill char, peat, and lignite. The carbon is produced by charring the raw material in the absence of air below 600°C, followed by an activation step consisting of partial oxidation. Carbon dioxide may be employed as an oxidizing agent at 600-700°C (Reaction 18) or the carbon may be oxidized by water at 800-900°C.

$$CO_2 + C \rightarrow 2CO \quad \text{...(18)}$$

$$H_2O + C \rightarrow H_2 + CO \quad \text{...(19)}$$

These processes develop porosity, increase the surface area, and leave the C atoms in arrangements that have affinities for organic compounds.

A major reason for the effectiveness of activated carbon as an adsorbent is its tremendous surface area. A solid cubic foot of carbon particles may have a combined pore and surface area of approximately 10 square miles.

Removal of organics may also be accomplished by absorbent synthetic polymers. Such polymers as Amberlite XAD-4 have hydrophobic surfaces and strongly attract relatively insoluble organic compounds, such as chlorinated pesticides. Oxidation of dissolved organics holds some promise for their removal. Ozone, hydrogen peroxide, molecular oxygen (with or without catalysts), chlorine and its derivatives, permanganate, or ferrate can be used.

Removal of Dissolved Inorganics

In order for complete water recycling to be feasible, inorganic solute removal is essential. The effluent form secondary waste treatment generally contains 300-400 mg/L more dissolved inorganic material than does the municipal water supply. It is obvious, therefore, that 100% water recycle without removal of inorganics would cause the accumulation of an intolerable level of dissolved material. Even when water is not destined for immediate reuse, the removal of the inorganic nutrients phosphorus and nitrogen is highly desirable to reduce eutrophication downstream. In some cases, the removal of toxic trace metals is needed.

One means for removing inorganics from water is distillation, although the energy required is generally too high for the process to be economically feasible, and volatile substances, such as ammonia and odorous compounds, may be carried into the product. Freezing produces a very pure water, but is considered uneconomical with present technology. Membrane processes considered most promising for bulk removal of inorganics from water are electrodialysis and reverse osmosis. Ion exchange is also effective for inorganics removal.

Ion Exchange

The ion exchange process used for removal of inorganic salts, represented below as dissolved ions of M^+ and X^-, consists of passing the water successively over a solid cation exchanger and a solid anion exchanger so that the ions are replaced by water:

$$H^{+-}\{Cat(s)\} + M^+ + X^- \rightarrow M^{+-}\{Cat(s)\} + H^+ + X^- \quad ...(20)$$

$$OH^{-+}\{An(s)\} + H^+ + X^- \rightarrow X^{-+}\{An(s)\} + H_2O \quad ...(21)$$

In these reactions $^-\{Cat(s)\}$ represents the solid cation exchanger and the notation $OH^{-+}\{An(s)\}$ represents the solid anion exchanger. The cation exchanger is regenerated with strong acid and the anion exchanger with strong base.

Demineralization by ion exchange generally produces water of a very high quality. Unfortunately, some organic compounds in wastewater foul ion exchangers, and microbial growth on the exchangers can diminish their efficiency. In addition, regeneration of the resins is expensive, and the concentrated wastes from regeneration require disposal in a manner that will not damage the environment.

Reverse Osmosis

Reverse osmosis is a very useful technique for the removal of dissolved inorganics from water. Basically, reverse osmosis consists

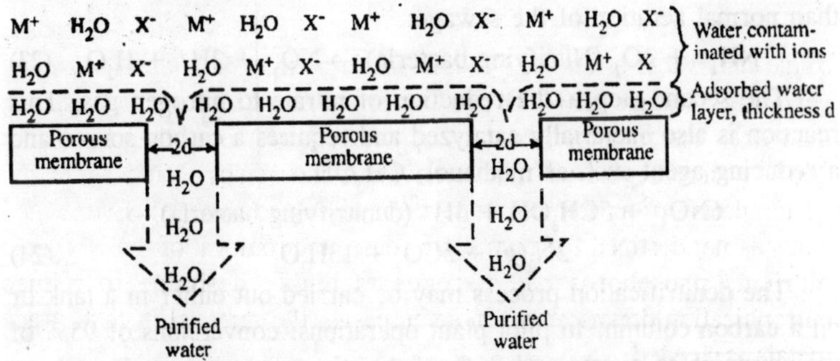

Fig. 6.4. Solute removal from water by reverse osmosis.

of forcing pure water through a semipermeable membrane that allows the passage of water but not of other material. This process depends on the preferential sorption of on the surface of the membrane, which is composed of porous cellulose acetate or polyamide. Pure water from the sorbed layer is forced through pores in the membrane under pressure.

Phosphorus Removal

Advanced waste treatment normally requires removal of phosphorus to reduce algal growth. Municipal wastes typically contain approximately 25 mg/L of phosphate (as orthophosphates, polyphosphates, and insoluble phosphates), and the efficiency of phosphate removal must be quite high to prevent algal growth. Normally, the activated sludge process removes about 20% of the phosphorus form sewage.

Chemically, phosphate is most commonly removed by precipitation. Lime, $Ca(OH)_2$, is the chemical usually used to precipitate phosphorus:

$$5Ca(OH)_2 + 3HPO_4^{2-} \rightarrow Ca_5OH(PO_4)_3(s) + 3H_2O + 6OH^- \quad ...(22)$$

Nitrogen Removal

Next to phosphorus, nitrogen is the algal nutrient most commonly removed as part of advanced wastewater treatment. Nitrogen in municipal wastewater generally is present as organic nitrogen or ammonia. Ammonia is the primary nitrogen produce produced by most biological waste treatment processes. It may be stripped in the form of NH_3 gas from the water by air after the pH has been raised to approximately 11.5 by the addition of lime (which also serves to remove phosphate).

Nitrification followed by denitrification is a promising technique for the removal of nitrogen from wastewater. The first step is an essentially complete conversion of ammonia and organic nitrogen to nitrate under strongly aerobic conditions, achieved by more extensive than normal aeration of the sewage:

$$NH_4^+ + 2O_2 \text{ (Nitrifying bacteria)} \rightarrow NO_3^- + 2H^+ + H_2O \quad ...(23)$$

The second step is the reduction of nitrate to nitrogen gas. This reaction is also bacterially catalyzed and requires a carbon source and a reducing agent such as methanol, CH_3OH.

$$6NO_3^- + 5CH_3OH + 6H^+ \text{ (denitrifying bacteria)} \rightarrow 3N_2(g) + 5CO_2 + 13H_2O \quad ...(24)$$

The denitrification process may be carried out either in a tank or on a carbon column. In pilot plant operations, conversions of 95% of the ammonia to nitrate and 86% of the nitrate to nitrogen have been achieved.

SLUDGE

Perhaps the most pressing water treatment problem at this time has to do with sludge collected or produced during water treatment. Finding a safe place to put the sludge or a use for it has proven troublesome, and the problem is aggravated by the growing number of water treatment systems. Improper disposal of wastes continues to be a subject of public and governmental concern.

Some sludge is present in wastewater and may be collected from it. Such sludge includes human wastes, garbage grindings, organic wastes and inorganic silt and grit from storm water runoff, and organic and inorganic wastes from commercial and industrial sources. There are two major kinds of sludge generated in a waste treatment plant. The first of these is organic sludge from activated sludge, trickling filter, or rotating biological reactors. The second is inorganic sludge from the addition of chemicals, such as in phosphorus removal described in the preceding section.

Most commonly, sewage sludge is subjected to anaerobic digestion in a digester designed to allow bacterial action to occur in the absence of air. This reduces the mass and volume of sludge and ideally results in the formation of a stabilized humus. Disease agents are also destroyed in the process.

Following digestion, sludge is generally conditioned and thickened to concentrate and stabilize it and make it more dewaterable. Relatively inexpensive processes, such as gravity thickening, may be employed to get the moisture content down to about 95%. Sludge may be further conditioned chemically by the addition of iron or aluminum salts, lime, or polymers.

Sludge dewatering is employed to convert the sludge from an essentially liquid material to a damp solid containing not more than about 85% water. This may be accomplished on sludge drying beds consisting of layers of sand and gravel. Mechanical devices may also be employed, including vacuum filtration, centrifugation, and filter presses. Heat may be used to aid the drying process.

Some of the alternatives for the ultimate disposal of sludge include land spreading, ocean dumping, and incineration. Each of these choices has disadvantages, such as the presence of toxic substances in sludge spread on land or the high fuel cost of incineration.

Rich in nutrients, waste sewage sludge contains around 5% N, 3% P, and 0.5% K on a dry-weight basis and can be used to fertilize and condition soil. The humic material in the sludge improves the physical properties and cation-exchange capacity of the soil. Among the factors limiting this application of sludge are excess nitrogen pollution of runoff water and groundwater, survival of pathogens, and the presence of heavy metals in the sludge.

A variety of chemical sludges are produced by various water treatment and industrial processes. Among the most abundant of such sludge in alum sludge produced by the hydrolysis of Al(III) salts used in the treatment of water, which creates gelatinous aluminum hydroxide:

$$Al^{3+} + 3OH^{-}(s) \rightarrow Al(OH)_3(aq) \qquad ...(25)$$

Alum sludges normally are 98% or more water and are very difficult to dewater.

Both iron(II)) and iron(III) compounds are used for the precipitation of impurities from wastewater via the precipitation of $Fe(OH)_3$. The sludge contains $Fe(OH)_3$ in the form of soft, fluffy precipitates that are difficult to dewater beyond 10 or 12% solids.

The addition of lime, $Ca(OH)_2$, or quicklime, CaO, to water is used to raise the pH to about 11.5 and causes the precipitation of $CaCO_3$, along with metal hydroxides and phosphates. Calcium carbonate is readily recovered from lime sludges and can be recalcined to produce CaO, which can be recycled through the system.

Metal hydroxide sludges are produced in the removal of metals such as lead, chromium, nickel, and zinc from wastewater by rasing the pH to such a level that the corresponding hydroxides or hydrated metal oxides are precipitated. The disposal of these sludge is a substantial problem because of their toxic heavy metal content. Reclamation of the meals is an attractive alternative for these sludges.

Pathogenic (disease-causing) microorganisms may persist in the sludge left from the treatment of sewage. Many of these organism present potential health hazards and there is risk of public exposure when the sludge is applied to soil. The most significant organisms in municipal sewage sludge include the following: (1) indicators of pollution, including fecal and total coliform; (2) pathogenic bacteria including *Salmonellae* and *Shigellae;* (3) enteric (intestinal) viruses, including enterovirus and poliovirus; and (4) parasites, such as *Entamoeba histolytica* and *Ascaris lumbricoides*. Therefore, it is necessary both to be aware of pathogenic microorganisms in municipal wastewater treatment sludge and to find a means of reducing the hazards caused by their presence.

Several ways are recommended to significantly reduce levels of pathogens in sewage sludge. These include prolonged aerobic digestion, air drying, composting, or lime stabilization in which sufficient lime is added to raise the pH of the sludge to 12 or higher.

WATER DISINFECTION

Chlorine is the most commonly used disinfectant employed for killing bacteria in water. Added to water, chlorine rapidly hydrolyzes according to the reaction

$$Cl_2 + H_2O \rightarrow H^+ + Cl^- + HOCl \qquad ...(26)$$

where HOCl is hypochlorous acid. Sometimes, hypochlorite salts are substituted for chlorine gas as a disinfectant. Calcium hypochlorite, $Ca(OCl)_2$, is commonly used. The hypochlorites are safer to handle than gaseous chlorine.

The two chemical species formed by chlorine in water, HOCl and OCl^- are known as *free available chlorine*. Free available chlorine is very effective in killing bacteria. In the presence of ammonia, HOCl reacts with NH_4^+ ion to form monochloramine (NH_2Cl), dichloramine ($NHCl_2$), and trichloramine (NCl_3). The chloramines are called *combined available chlorine*. Chlorination practice frequently provides for formation of combined available chlorine which, although a weaker disinfectant than free available chlorine, is more readily retained as a disinfectant throughout the water distribution system.

Chlorine Dioxide

Chlorine dioxide, ClO_2, is an effective water disinfectant that is of particular interest because, in the absence of impurity Cl_2, it does not produce impurity trihalomethanes in water treatment. In acidic and neutral water, respectively, the two half-reactions for ClO_2 acting as an oxidant are the following:

$$ClO_2 + 4H^+ + 5e \leftarrow\rightarrow Cl^- + 2H_2O \qquad ...(27)$$

$$ClO_2 + e^- \leftarrow\rightarrow ClO_2^- \qquad ...(28)$$

In the neutral pH range, chlorine dioxide in water remains largely as molecular ClO_2 until it contacts a reducing agent with which to react. Chlorine dioxide is a gas that is violently reactive with organic matter and explosive when exposed to light. For these reasons, ClO_2 is not shipped, but is generated on-site by processes such as the reaction of chlorine gas with solid sodium hypochlorite:

$$2NaClO_2(s) + Cl_2(g) \leftarrow\rightarrow 2ClO_2(g) + 2NaCl(s) \qquad ...(29)$$

A high content of elemental chlorine in the product may require its purification to prevent unwanted side-reactions from Cl_2.

As a water disinfectant, chlorine dioxide does not chlorinate or oxidize ammonia or other nitrogen-containing compounds. Some concern has been raised over possible health effects of its main degradation by-products, ClO_2^- and ClO_3^-.

Ozone

Ozone is sometimes used as a disinfectant in place of chlorine, particularly in Europe. Basically, air is filtered, cooled, dried, and pressurized, then subjected to an electrical discharge of approximately 20,000 volts. The ozone produced is then pumped into a contact chamber where water contacts the ozone for 10-15 minutes. Concern over possible production of toxic organochlorine compounds by water chlorination processes has increased interest in ozonation. Furthermore, ozone is more destructive to viruses than is chlorine, Unfortunately, the solubility of ozone in water is relatively low, which limits its disinfective power.

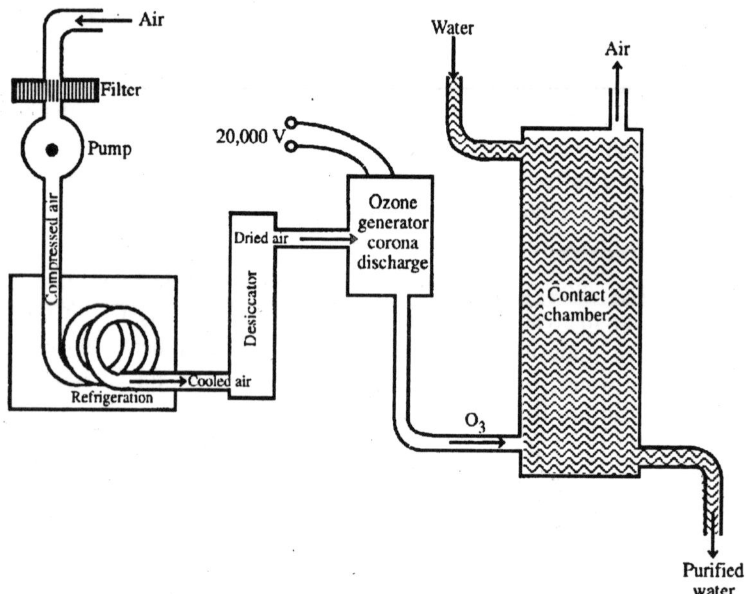

Fig. 6.5. A schematic diagram of a typical ozone water treatment system.

Natural Water Purification Processes

Virtually all the material that waste treatment processes are designed to eliminate may be absorbed by soil or degraded in soil; in fact, most are essential for soil fertility. The most basic thing that wastewater provides to plants is the water that is essential to plant growth. In addition to water, wastewater is normally very rich in essential plant nutrients, which are usually provided by fertilizers.

These nutrients include phosphorus, largely as inorganic phosphate ions ($H_2PO_4^-$ and HPO_4^{2-}) ions), nitrogen (primarily as ammonium nitrogen, NH_4^+), and potassium (K^+ ion). Wastewater may also contain organically bound phosphorous and nitrogen that undergo biologically-mediated mineralization processes to provide the inorganic forms of these elements required by plants. Wastewater also contains essential trace elements and vitamins. Stretching the point a bit, the degradation of organic wastes provides the CO_2 essential for photosynthetic production of plant biomass.

Soil may be viewed as a natural filter for wastes. Most organic matter is readily degraded in soil and, in principle, soil constitutes an excellent treatment system for water. Soil can function to provide primary, secondary, and tertiary treatment of wastewater and has the additional advantage of not requiring expensive disposal of sludge after treatment. Soil has physical, chemical, and biological characteristics that can enable wastewater detoxification, biodegradation, chemical decomposition, and physical and chemical fixation. Soil is a natural medium for a number of living organisms that may have an effect upon biodegradation of wastewaters, including those that contain industrial wastes. Of these, the most important are bacteria, including some kinds of *Agrobacterium*, *Arthrobacteri*, *Bacillus*, *Flavobacterium*, and *Pseudomonas*. Actinomycetes and fungi are important in decay or vegetable matter and may be involved in biodegradation of wastes. Other unicellular organisms that may be present in or on soil are protozoa and algae. Soil animals, such as earthworms, affect soil parameters such as soil texture. The growth of plants in soil may have an influence on its waste treatment potential in such aspects as uptake of soluble wastes and erosion control.

Early civilizations, such as the Chinese, used human organic wastes to increase soil fertility, and the practice continues today. The ability of soil to purify water was noted well over a century ago. In 1850 and 1852, J. Thomas Way, a consulting chemist to the Royal Agricultural Society in England, presented two papers to the Society entitled "Power of Soils to Absorb Manure." Mr. Way's experiments showed that soil is an ion exchanger. Much practical and theoretical information on the ion exchange process resulted from his work.

Experiments involving the direct application of wastewater to soil have shown appreciable increases in soil productivity. A process called *overland flow* has been used in which wastewater containing pulverized solids is allowed to trickle over sloping soil. Suspended solids, BOD, and nutrients are largely removed. This method is most applicable to rural communities in relatively warm climates. Approximately one acre of land is required to handle the sewage from 200 persons.

If such systems are not properly designed and operated, odor can become an overpowering problem. The author of this book is remained of driving into a small town, recalled from some years before as a very pleasant place, and being assaulted with a virtually intolerable odor. The disgruntled residents pointed to a large spray irrigation system on a field in the distance—unfortunately upwind—spraying liquefied pig manure as part of an experimental feedlot waste treatment operations. The experiment was not deemed a success and was discontinued by the investigators, presumably before they met with violence from the local residents.

Industrial Wastewater Treatment by Soil

Wastes that are amenable to land treatment are biodegradable organic substances, particularly those contained in municipal sewage and in wastewater from some industrial operations, such as food processing. However, through acclimation over a long period of time, soil bacterial cultures may develop that are effective in degrading normally recalcitrant compounds that occur in industrial wastewater. Acclimated microorganisms are found particularly at contaminated sites, such as those where soil has been exposed to crude oil for many years.

Land treatment is most used for petroleum refining wastes and is applicable to the treatment of fuels and wastes from leaking underground storage tanks. It can also be applied to biodegradable organic chemical wastes, including some organohalide compounds. Land treatment is not suitable for the treatment of wastes containing acids, bases, toxic inorganic compounds, salts, heavy metals, and organic compounds that are excessively soluble, volatile, or flammable.

7

SEVAGE TREATMENT

Waste is any movable material that is perceived to be of no further use and that is permanently discarded. Once in the environment, wastes frequently cause damage to ecosystems and/or human health and therefore act as pollutants.

Successful waste management can largely avoid such pollution. This chapter introduces the more widely available strategies and technologies that can be effective in this area. The first three sections deal with the approaches used in the management of the relatively low-hazard wastes that are generated in bulk by industrial, commercial and domestic activity. Consideration of the options available for the safe treatment and disposal of high-hazard wastes is given in the fourth section. The chapter closes with a brief introduction to the concepts of waste minimisation, cleaner production and integrated waste management. If more widely adopted, these ideas have the potential to greatly improve current waste management practices.

WASTES FROM FOSSIL FUEL COMBUSTION

The main wastes generated during the combustion of fossil fuels are sulfur dioxide, NO_N, carbon monoxide, unburnt hydrocarbons, particulates, residual solids (including ash) and carbon dioxide. The technologies that are available for the management of these wastes are briefly reviewed in this section.

Sulfur Dioxide

Fossil fuels contain both organic sulfur (e.g. in thiophene rings) and inorganic sulfides (principally H_2S in natural gas and FeS_2 in coal). During combustion these react with atmospheric oxygen (O_2) to produce sulfur dioxide (SO_2).

The sulfur content of fossil fuels varies considerably. For example, coals and fuel oils generally contain 1-4%, and 3-4% S respectively. However, there are naturally occurring low-sulfur fuels (e.g., coals <1% S and fuel oils <0.5% S). Clearly, burning these preferentially is one of the options available for diminishing the emissions of sulfur dioxide. Unfortunately, this is of only limited applicability as supplies of these low-sulfur fuels are comparatively small.

Another alternative is the dilution and dispersion of the sulfur dioxide produced, principally by building taller chimneys. This has found favour in the past and has had noticeable success in the reduction of local levels of pollution. Unfortunately, it has had no impact on overall contamination; in effect, 'what goes up must come down'.

Fuel cleaning processes that remove sulfur are routinely applied to natural gas, oil and coal. These are now considered in turn.

Natural gas contains variable amounts of hydrogen sulfide (H_2S). This may be effectively removed by a number of processes including adsorption onto zeolites (a type of aluminosilicate mineral). The hydrogen sulfide may then be oxidised *in situ* with hot sulfur dioxide to yield sulfur vapour and regenerated zeolite adsorbant. The sulfur is then condensed and sold, while the zeolite is reused.

The desulfurisation of oils is desirable for a number of reasons that are unrelated to the lowering of sulfur dioxide emissions. These include avoiding the deactivation (poisoning) of platinum catalysts used during oil processing. Consequently, oil desulfurisation was practised before the environmental need to reduce sulfur dioxide emissions was recognised. The main process involved is hydrodesulfurisation. During this the oil is reacted with hydrogen (H_2) at elevated temperatures, under pressure and in the presence of a catalyst. This converts the sulfur to hydrogen sulfide which can then be separated as a gas.

An important consequence of oil desulfurisation is that motor spirit (petrol, gasoline) has a very low sulfur content (between 0.026% in US Premium grade and 0.040% in the UK). As a result, transport makes very little contribution to the total anthropogenic emissions of sulfur dioxide.

Coal is cleaned by the separation of the organic fuel from the inorganic ash-forming mineral impurities that it contains. This may be done on the basis of density, for the fuel has a lower specific gravity (1.1 to 1.8) than the impurities (from about 2 to about 5). In one process the raw coal is finely ground, so that most of the mineral particles become distinct from the fuel. The ground raw coal is then agitated in a mixture of air, water, oil and surfactant. The denser particles sink, while the others are held by surface tension at the interface between the liquid and the air. The cleaned coal is then isolated in a settling tank, where the air/oil/water mixture is allowed to separate, causing the fuel to sink.

Processes such as this may remove much of the inorganic sulfur fraction (FeS_2 has a specific gravity of 4.5). In a typical British coal, about half of the sulfur is inorganic, the rest forming part of the organic matrix of the fuel. It is now technologically possible to remove some of this also, though it is currently not economically viable to do so.

Vast amounts of coal are consumed worldwide, particularly during the production of electricity. This, coupled with the relative inefficiency of the routine coal cleaning process, makes this fuel by far the largest single contributor to anthropogenic sulfur dioxide emissions. There is therefore considerable interest in the removal of sulfur dioxide prior to the release of the flue gases.

Sulfur dioxide removal rates of 90% can be achieved from the combustion zone in boilers that are based on *fluidised bed combustion* (FBC) technology. In such systems the fuel is added in a pulverised form to a bed of inert material (e.g. sand or coal ash). This is kept in a state of agitation (i.e. fluidised) by a strong updraught of air, which acts as the oxidant. Such systems allow the coal to be burnt efficiently at relatively low temperatures ('900°C).

As an alternative, sulfur dioxide can be removed downstream of the boiler after the fly ash has been removed, a process called flue gas desulfurisation (FGD). FGD can be highly efficient: 90% removal rates are generally achievable. In a typical system, an aqueous slurry of an alkaline absorbant, commonly lime, or limestone, is passed in a fine spray through the flue gases. Sulfites and sulfates are therefore generated during this 'scrubbing' process:

$$Ca(OH)_2 + SO_2 \rightarrow CaSO_3 + H_2O$$

$$CaCO_3 + SO_2 \rightarrow CaSO_3 + CO_2$$

$$CaSO_3 + 1O_2 \rightarrow CaSO_4$$

The last of these reactions can be encouraged by the injection of air into the sump of the scrubbing tower. This yields high-quality gypsum ($CaS0_4.2H_2O$) which can be sold for use in plasterboard and other building materials.

NO_x, Carbon Monoxide and Unburnt Hydrocarbons

The burning of fossil fuels in air produces nitric oxide (NO) and, to a lesser extent, nitrogen dioxide (NO_2); these are collectively known as NO,. They are formed by the reaction of atmospheric oxygen with nitrogen at the high temperatures reached during combustion. The nitrogen may originate from either the air or the fuel, thereby producing thermal-NO_x and fuel-NO_x respectively.

The problem of fuel-NO_x is primarily associated with coal because it has relatively high levels of nitrogen (1-2%) compared with other fossil fuels. For example, natural gas is virtually nitrogen-free, while fuel oil contains <0.5% N.

Clearly, thermal-NO, formation occurs whenever fuels are burnt in air. This allows transport to be a major contributor to NO_x emissions. For example, in the UK about half of NO_x is traffic-related. The remainder originates from stationary producers, particularly electricity generating stations.

Reduction in the emissions of NO_x can be achieved by alterations to the combustion process. The reactions that produce thermal-NO_x are endothermic and are therefore favoured by high temperatures. Lowering the temperature of combustion by, for example, recycling exhaust gases will therefore diminish NO_x emissions. Unfortunately, this will also reduce the Carnot efficiency of any heat-to-work device driven by the fuel. If used in a motor vehicle, NO_x reduction by this method will therefore be at the expense of fuel economy.

Fuel-NO_x emissions can also be controlled by adjustments to the combustion process. Fuel nitrogen that has been oxidised to nitric oxide may then be reduced to molecular nitrogen by either fuel-derived volatiles or char, for example:

$$2NO_{(g)} + 2CO_{(g)} \rightarrow N_{2(g)} + 2CO_{2(g)}$$
$$2NO_{(g)} + 2C_{(s)} \rightarrow N_{2(g)} + 2CO_{(g)}$$

These reactions can be encouraged by allowing the early stages of the combustion process to be carried out under fuel-rich conditions, followed by an injection of air into the flame when it is more mature, allowing the char to be oxidised. This approach, called staged combustion, when used alone can result in the removal of up to 50% of NO_x in coal-fired stations.

The treatment of flue gases can also lead to NO_x removal. The approach used is dependent on whether the source is static or mobile. In the former case, either ammonia, NH_3 (with or without a catalyst), or urea, $(NH_2)_2CO$, is injected into the stack gases, causing the NO_x to be reduced:

$$6NO + 4NH_3 \rightarrow 5N_2 + 6H_2O$$

and

$$4NO + 4NH_3 + O_2 \rightarrow 4N_2 + 6H_2O$$

or

$$2(NH_2)_2CO + 6NO \rightarrow 5N_2 + 2CO_2 + 4H_2O$$

The treatment of vehicular emissions may be achieved by the catalytic reduction of NO_x to molecular nitrogen at the expense of carbon monoxide (CO) present in the exhaust gases:

$$2NO + 2CO \xrightarrow[\text{on an inert support)}]{\text{catalyst (e.g. rhodium}} 2CO_2 + N_2$$

Then, air may be injected and the gases allowed to pass over an oxidation catalyst such as platinum or palladium on an inert support. This will facilitate the conversion of any residual carbon monoxide to carbon dioxide and any unburnt hydrocarbons present in the waste stream to carbon dioxide and water.

Particulates

Both stationary sources and Diesel-powered vehicles produce significant amounts of particulates. Where attempted, the recovery of these contaminants from the stack gases of the former source is generally very successful. The technologies used are based on cyclones, electrostatic precipitators and/or fabric filters (bag filters).

During the operation of a cyclone, the exhaust gases enter the top of its essentially cylindrical body, at a tangent. This causes them to move downwards in a helical fashion, generating centripetal forces that drive the particulates to the walls, from where they fall, exiting the cyclone at the bottom. The cleaned gases then leave the top of the cyclone via the pipe at its centre.

Electrostatic precipitators (ESPs) operate by virtue of a potential difference of 30 to 60 kV between the wires and plates that they contain. This causes a very steep gradient in the electric field around the wires and a concomitant high concentration of ions. These charge the particles of the effluent stream, which are then accelerated towards the plates by the potential difference. The dust may then be dislodged from the plates by agitation, allowing it to fall into a collection hopper.

Fabric filters (bag filters) physically remove particulates from the exhaust gases that are made to pass through them. The filters may be of many designs, although tubular constructions are common. The dust burden is periodically removed by either mechanical shaking and/or the reversal of the direction of gas flow. There is increasing concern over the sooty particulates from Diesel engines, as epidemiological studies indicate that these contaminants may cause a range of health problems including heart disease. Currently, in the UK, Diesel engines account for about 40% of all black smoke emissions. It seems likely that this percentage will go up as the popularity of Dieselpowered vehicles increases.

Control of Diesel particulates is technologically difficult, though two approaches seem promising. The first involves improving the homogeneity of the fuelair mixture at the time of firing, so ensuring a more complete burn. The second relies on ceramic filters that may be cleaned either physically, by compressed air, or chemically, by heating in the presence of air.

Residual solids

The combustion of finely ground coal in electricity generating stations produces very large quantities of residual solids. These are ashes and, more recently, the products of limestone-based desulfurisation.

Two types of ash are generated, namely pulverised fuel ash (PFA) and furnace bottom ash (FBA); together these amount to about 12-13 Mte a-' in England and Wales alone. PFA is collected as a particulate from the flue gases and accounts for 80% of the total. Both of these products are used in cementitious materials. Despite this, in areas where production outstrips demand, considerable quantities are sent to landfill.

As previously mentioned, lime- or limestone-based desulfurisation post PFA removal can yield highquality gypsum ($CaSO_4.2H_2O$). Clearly this has commercial value. However, the vast amounts produced may be sufficient to swamp the market, necessitating other disposal routes including landfill.

Carbon Dioxide

All fossil fuel combustion leads to the generation of carbon dioxide. Many exotic means of diminishing the contamination of the atmosphere with this gas have been suggested. Included amongst these is the possibility of increasing the primary productivity of the oceans. It is thought that this may be achieved by adding relatively small amounts of iron to areas that are deficient in this element. According to this hypothesis, the consequent increased rates of photosynthesis will result in the absorption of carbon dioxide. Recent large-scale experiments in the Pacific demonstrated that a single addition of iron salts did indeed promote productivity, at least in the short term. However, a fully concomitant net consumption of carbon dioxide did not occur. One possible explanation of this is that an increased biomass of photosynthetic plankton encouraged the activity of grazing zooplankton, and that the respiratory activity of these organisms recycled a large proportion of the carbon dioxide originally absorbed. It seems doubtful that the seeding of oceans with iron represents a feasible means of controlling atmospheric levels of carbon dioxide. A more practicable approach is to look for improved fuel efficiency. This would be of even greater efficacy if coupled with a switch to lowcarbon fuels such as methane, or even non-carbon fuels including hydrogen (H_2), which may be generated by hydroelectric power.

Low-hazard Solid Wastes

Solid wastes (refuse) may be categorised by source into mining, agricultural, industrial and urban (municipal) waste. The last of these includes wastes generated by commerce, local authorities and domestic households.

On a global basis, data concerning the amounts of solid waste generated are inadequate. The problem of insufficient data is compounded by variations in the definition of waste from country to country, making comparisons difficult. However, it is clear that the problem is enormous. For example, for the period of the late 1980s it has been estimated that the OECD countries generated in excess of 1.85×10^{12} kg of solid waste per year. What is more, in some respects the situation appears to be getting worse, particularly in the developing countries. For example, in the moredeveloped world, *municipal* solid waste generation increased from about 3.2×10^{11} kg a^{-1} in 1970 to 4×10^{11} kg a^{-1} in 1990 (~25%). During the same period, the production of refuse in the developing nations underwent an even more rapid rate of increase from 1.6×10^{11} kg a^{-1} to 3.2×10^{11} kg a^{-1} (~100%).

Most solid waste is of low intrinsic hazard. Nonetheless, if mismanaged even this has the potential to cause a diversity of problems, ranging from aesthetic deterioration of the environment through to significant increases in the incidence of disease and the pollution of drinking waters.

Of economic necessity, mining waste is usually disposed of on land near to the mine workings, often forming large spoil heaps. Agricultural solid waste, including crop residues and dung, have fertiliser and soil-conditioning value. Therefore, they are generally disposed of *in situ*. This leaves industrial and urban wastes to consider. The main disposal options for low-hazard waste from these sources are, in approximate order of increasing desirability: indiscriminate dumping, landfill (organised dumping on land), incineration (if organic) and reuse.

Indiscriminate dumping is an almost ubiquitous problem. However, it is particularly acute in many of the cities of the developing world. This is despite the relatively low *per capita* generation of domestic refuse in these cities (~145-330 kg a^{-1}) compared with the production rate in the cities of the more-developed countries (~255-655 kg a^{-1}). The main cause of the problem is that, in the less-developed nations, only about 50-70% of urban solid waste is collected. The remainder accumulates in the streets and open spaces, where it becomes a breeding ground for vermin, spreading disease. Where collection does occur, it frequently results in the formation of open tips that support large numbers of waste-pickers who derive an income from the reusable articles that have beep. discarded.

Unlike open tips, properly managed landfill sites are a very effective means of low-hazard solid waste disposal. If the waste is covered with soil on a daily basis, odour release is controlled and vermin are discouraged. Under these conditions, these facilities are called 'sanitary landfill' sites and need not be a source of either public nuisance or health hazard. Once full, anaerobic degradation of the material within the capped landfill occurs over a 3-10 year period. During this time the site is of little use as the ground settles and gas is evolved. This is mainly carbon dioxide and methane, controlled removal of which is desirable as this avoids the danger of explosions.

The incineration of low-hazard solid waste with high organic contents in large purpose-built facilities is attractive for several reasons. Principal among these is the considerable reduction in the volume of solid material achieved by this process. In the case of domestic refuse this is generally about 75%. What is more, the residue does not undergo anaerobic digestion when placed in landfill; consequently, settlement and gas generation do not occur, allowing the site, once full, to be built upon.

The major drawback of incineration is the generation of flue gases and particulates. These can be minimised by the application of technologies that are essentially the same as those used to clean the stack gases of static fossil fuel burning facilities.

Solid wastes sent for incineration frequently include chlorine-containing organic substances, such as polyvinylchloride (PVC). The burning of these leads to the formation of trace amounts of polychlorinated dibenzo-p-dioxins and polychlorinated dibenzofurans (PCDDs and PCDFs), some of which are highly toxic. The presence of these materials in the emissions and ashes produced by incinerators of all kinds has generated considerable opposition to this type of waste treatment facility.

There are several ways in which low-hazard solid waste can be reused. The main processes available are the recycling of individual materials, the generation of refuse-derived fuel, composting, and thermochemical treatment.

Urban refuse from an industrialised country may be expected to have a composition. Virtually all of the components listed would have a reuse value, if they were collected separately. Unfortunately, this is largely impracticable and economically unviable. However, in recent years there has been a move towards the separate collection of the more valuable items, particularly paper, glass, aluminium, steel, plastics and fabric. In the UK this has been achieved largely by the willingness of the public to take these items to specialised receptacles ('banks'), often situated in car parks or at household waste disposal facilities.

An alternative approach is to separate the mixed waste after collection. In the case of the ferrous metals this is readily achieved by magnetic means. While the separation of the other components is more difficult, it can be achieved, to some extent, on the basis of density.

The recycling of waste has other environmental benefits besides those directly associated with direct waste reduction. Waste recycling generally consumes fewer resources and produces less pollution than the winning of materials from virgin sources.

There is evidence to suggest that the recycling of some solid wastes is becoming more significant. For example, on a worldwide basis, in 1971 recycled aluminium formed about 16% of the total yearly consumption of this metal; by 1987 this had grown to over 23%. What is more, during the same period, total consumption of aluminium increased from about 1.2×10^{10} to approximately 2.2×10^{10} kg a^{-1}.

The success of recycling activity varies considerably from one country to another. For example, while the UK recycles 14% of its glass, the Netherlands recycles 62%. Ironically, because of the activities of waste-pickers, lessdeveloped countries frequently have high rates of refuse reuse. This is despite the generally lower levels of valuable material contained in the solid waste of these counties.

Density-based solid waste separation can generate an inorganic fraction (containing metals and glass for recycling) and an organic fraction of sufficiently high calorific value to use as a fuel. This refuse-derived fuel may be used in either a shredded or a pelletised form. Recently, waste tyres have been used to fire cement kilns; one advantage of this process is that the ash becomes integrated with the product, obviating the need for its disposal.

Composting of refuse, under aerobic conditions, offers another way of producing a useful product. The material generated finds use as a soil conditioner and low-grade fertiliser.

Finally, solid waste can be degraded by a variety of thermochemical processes, including pyrolysis (i.e. chemical breakdown achieved by heating in an anaerobic atmosphere). The product of most of these processes is a solid residue together with fuel gas and oil. The attraction of such processes is that they simultaneously reduce the mass of the solid that has to be disposed of, while producing fuels with good handling properties.

Low-hazard Waste Waters (Sewage)

The water within the sewerage system of a community is called sewage. It consists of the outflow from domestic and industrial premises and, in some cases, the run-off from roads. This waste water is usually greatly diluted by the ingress of ground water through leaking pipe joints.

Table 7.1. The composition of typical urban refuse derived from an industrialised country.

Component	*Proportion/% by weight*
Paper	35
Garden waste	16
Food waste	15
Metals	10
Glass	10
Plastics, rubber and leather	7
Rags	2
Miscellaneous	5

Only a very small fraction of sewage (0.05%) is waste material, the rest being water. Despite its apparently low waste content, the discharge of untreated sewage into surface waters can lead to gross pollution.

Sewage treatment is primarily aimed at lowering the pathogen content of the waste. Additional objectives include a decrease in its biochemical oxygen demand and solids content. These objectives may be achieved in a series of stages.

Preliminary treatment removes the larger objects within the raw sewage (lumps of wood, bottles, sanitary towels, toilet paper, etc.) and the grit. This may be the only treatment, if any, that is given to sewage prior to discharge in the sea. The processes used are mechanical. Screens constructed from iron bars remove the larger objects. In addition, macerators may be used to break up the more friable lumps so that they may proceed for further treatment. Grit separation is achieved using gravity under conditions where the less dense organic matter remains in suspension.

In all but the most rudimentary plants, the effluent from preliminary treatment is then subjected to primary treatment. During this process the sewage is allowed to slowly traverse a tank, allowing about half of the suspended solids to fall to the bottom. This produces primary sludge and settled sewage (also called primary effluent). The sludge is digested and the settled sewage then enters secondary treatment.

Secondary treatment is a biological process. Three designs of reactor are in common use, namely trickling (biological) filters, activated sludge tanks and oxidation (stabilisation) ponds. The last of these is only appropriate in warm climates.

Table 7.2. The potential savings of recycling.

	Potential saving/%			
	Aluminium	*Glass*	*Paper*	*Steel*
Water used	0	50	58	40
Energy used	90-97	4-32	23-74	47-74
Mining wastes	0	80	0	97
Polluting emissions to the atmosphere 95	20	74	85	
Polluting discharges to watercourses	97	0	35	76

A trickling filter is a tank filled with inert solid particles, typically in the size range 3.8-5.0 cm. These are covered with a mixed, essentially microbiological community, mainly developed from the sewage. Settled sewage is sprinkled over the top of the tank via moving booms. The sewage percolates down the filter, where its organic content is largely oxidised by organisms established on the solid support. Oxygen for this process is provided by air that is passively drawn into the tank.

In contrast, activated sludge tanks are actively aerated. In these the settled sewage is oxidised by a suspension of micro-organisms.

Both trickling filters and activated sludge tanks produce secondary sludge that is removed downstream in final settlement tanks. In the case of the activated sludge process most of this is returned to the aerated tank in order to maintain its biological community. The remainder may be added to the primary settlement tank, as may all of the sludge from trickling filters.

Oxidation ponds are large shallow (-1 m) tanks through which settled sewage may slowly pass. Microbial action releases nutrient species (CO_2, NH_3, NO_3) that sustain algal growth. The algae generate molecular oxygen during photosynthesis, which sustains the activity of the bacteria. An anaerobic sludge forms on the bottom of the ponds in which methane is produced. An advantage of oxidation ponds is that they can be harvested. The algae generated can be fed to animals or burnt. What is more, oxidation ponds can be used to raise fish, although there is a risk of pathogen transfer if the fish are used for human consumption.

The preliminary, primary and secondary treatments outlined above are highly successful. When in combination, they are capable of producing an effluent with less than 30 mg 1^{-1} of suspended solids and a BOD of less than 20 mg 1^{-1} (typical sewage contains 600mg1^{-1} of total solids, of which 200mgl^{-1} are suspended, and has a BOD of 3000mg1^{-1}). Pathogen populations are also greatly reduced. For example, the population of *Salmonella paratyphi* can be decreased by 84-99% by the use of trickling filters.

The final effluent from the treatment works is discharged into a river, lake or sea. Tertiary treatment is seen as desirable in locations where the degree of dilution of the effluent is small, or where potable water is to be withdrawn downstream for treatment and distribution. The purpose of this treatment is to further reduce the BOD and/or concentrations of suspended solids, nutrients, toxicants (such as heavy metals or poisonous organics) and/or pathogens. A wide range of technologies have been developed to facilitate the desired improvements, including oxidation ponds, sand filters, microstrainers, adsorption onto activated carbon, ion exchange, chemical precipitation, microfiltration, and disinfection. None of these, when operated alone, can bring about the desired reductions in all of the parameters listed above.

Conventional sewage treatment generates sludge from both the primary and secondary stages, to which may be added any sludge produced during tertiary treatment. These are mixed and digested in a two-stage process, the aim of which is to produce a material of reduced volume and acceptable odour that does not attract harmful insects or rodents. The first stage is anaerobic and is carried out at 27-35°C. It produces a gas that is approximately 72% methane and 28% carbon dioxide; this may be collected and used to generate heat (to warm the digester) and electricity.

The second stage is carried out in the open. The sludge is allowed to settle (thicken), producing the final product, digested sludge. Disposing of this is problematic. Options include dumping at sea, incineration, landfill, and use on land as a fertiliser/soil conditioner. The last of these is restricted by both transport costs and the heavy metal burden of the sludge. Landfill sites are increasingly scarce and dumping at sea is becoming restricted by legislation; in the US this practice was banned in 1991, and in the UK it will cease in 1998. Therefore the percentage of sludge that is sent for incineration appears likely to increase, at least in the more-developed countries.

For economic reasons, many households are not connected to mains sewerage systems. A commonly practised, but inferior, alternative is waste water treatment based on the septic tank. This acts as a combined sedimentation tank and anaerobic digester. The liquid effluent is allowed to soak away into the soil, while the sludge is periodically removed from the tank to be treated in a conventional sewage treatment plant.

Despite the existence of well-established technologies for the treatment and disposal of sewage, these are denied to many people. The situation appears to be getting worse; estimates indicate that by the year 2000 the number of people without sanitation facilities will reach 1880 million.

HIGH-HAZARD WASTES

Assessment of the amount of high-hazard waste generated on a global basis is problematic. This is in part because of inadequate record keeping, but also because there is no uniformly accepted definition of high-hazard waste. This makes international comparisons very difficult as some countries use definitions that are much more all-embracing than others. By way of illustration, it is interesting to note that 41% of the solid industrial waste generated in the USA is categorised as hazardous. This compares with 33.5% in Hungary, 3% in the UK and 0.3% in Japan and Italy. To a considerable degree, the enormous disparity between these figures reflects the differing regulatory frameworks within which the data were collected.

There are strong indications, however, that the production of high-hazard wastes is vast and expanding, both absolutely and as a proportion of industrial wastes as a whole. Global production of hazardous waste is estimated to be at least 3.38×10^{11} kg a^{-1}, about 80% of which is generated in the USA. In some countries the rate of increase appears to be phenomenal. For example, estimates of South Korean hazardous waste production for 1985 and 1989 are 1.2×10^{10} and 2.1×10^{10} kg a^{-1} respectively.

In addition to the hazardous wastes currently being produced, considerable amounts have been inappropriately disposed of in the past. Consequently, a large number of sites have been contaminated and are potentially hazardous. For example, 32000 such sites have been identified in the USA alone. The remedial treatment of these is likely to be extremely costly.

There is also a legacy of materials that are now known to be hazardous, but that were once in common usage. Disposal of these substances is likely to cause problems for some time to come. Notable amongst these are the polychlorinated biphenyls (PCBs), which found extensive use as dielectrics in transformers and asbestos, which was widely used as a building material.

For the purposes of the discussion here, highhazard wastes may be considered to be those that, when released in relatively small amounts, are capable of producing severe and/or long-lasting damage to human health or the environment. Included in our definition are materials that contain pathogens or radioactive isotopes, along with substances that are corrosive, toxic, flammable, violently reactive or explosive.

Treatment and Disposal

Strategies for the treatment of high-hazard wastes can be divided into those aimed at reuse, at destruction or at immobilisation.

Options for reuse include purification followed by recycling. This approach is frequently applied to solvents, as recovery of pure material from waste solvent is often achievable by distillation. An alternative approach is to use the waste from one process as a feedstock for another. For example, some waste oils may be mixed with fuel oils and burnt in industrial boilers.

There are instances where the waste from one process can be used to treat the waste from another. For example, prior to painting or electroplating, the oxide coat on steel is removed using acidic pickling liquors. Once spent, these may be reused as precipitating agents, removing phosphate from waste waters.

Reuse within the facility that generated the waste is desirable as the need for transport is minimised.

However, this is not always possible. In such cases, certain types of waste, including metals and solvents, may be passed on to commercial reclaimers. These then treat the wastes and sell them on as useful products. Alternatively, the wastes generated by one manufacturing company may be used directly by another. In some areas this has been encouraged by the establishment of 'waste exchanges'. These produce databases that list the wastes available in a given region, so facilitating trade in these commodities.

Destruction of high-hazard wastes is only applicable to those that are hazardous by virtue of molecules that they contain rather than their constituent elements. For example, the cyanide ion (CN) is found in wastes from metal processing industries. It is highly toxic even though both of its constituent elements are essential for life (Chapter 5). It is toxic by virtue of its affinity for the active sites of enzymes involved in respiration. Consequently ingestion of sufficient amounts of this ion results in rapid death. Wastes containing this species may be detoxified by treatment with chlorine (Cl_2), thus:

$$CN^-_{(aq)} + H_2O_{(l)} + Cl_{2(aq)} \rightarrow \underset{\text{cyanate}}{OCN^-_{(aq)}} + 2HCl_{(aq)}$$

followed by

$$OCN^-_{(aq)} + H_3O^+_{(aq)} \rightarrow NH_{3(aq)} + CO_{2(aq)}$$

Compare this process with the approach used in the treatment of waste waters containing toxic metals. These are generated by a number of industries including mining, metal-plating and ceramics manufacture. Unlike cyanide, metals cannot be destroyed. Consequently, treatment of waste streams contaminated with these elements involves the removal of metals from the aqueous phase. Clearly, the contaminants vary from source to source; however, they may include copper, nickel, cadmium, lead, chromium, mercury and/or zinc. These and many other heavy metals can be largely removed from the water by the addition of an anion that causes the precipitation of the metal as an insoluble salt. Anions used for this purpose include sulfate ($5O_4^{2-}$), sulfide (S^{2-}) and hydroxide (OH^-).

As in the case of cyanide destruction, discussed above, many of the destructive treatment methods are waste specific. However, there are others of more general applicability, notably thermo-chemical and biological processes.

The main thermo-chemical treatment used is incineration. In the case of wholly organic wastes, this method is also a means of complete waste disposal as the products are relatively harmless gases, principally $CO_{2(g)}$ and $H_2O_{(g)}$, which are subsequently vented into the air. The high temperatures reached also result in sterilisation. This is seen as a great advantage in the disposal of materials, such as hospital wastes, that may be contaminated with pathogens.

Incineration of high- and low-hazard wastes share the same drawback, namely potential air pollution. However, this may be minimised by a combination of:

1. high-temperature combustion (ideally $>1000^{\circ}C$);
2. a long residence time of the waste in a hot oxidising environment (>2 or 3 seconds, depending on the waste);
3. rapid stack gas cooling (to avoid the formation of toxic dioxins and furans: Box 16.3);
4. flue gas c1E aping.

By such means, modern plant is capable of achieving burnouts in excess of 99.99%.

Unfortunately, there are still a great number of incinerators that do not incorporate all of the features listed above. This has led to concern in recent years. In the UK, much of this has centred on the incineration of clinical waste as, until recently, this fell outside the reach of all environmental law.

The incorporation of organic high-hazard wastes into the input stream of cement kilns has been used for many years as an ultimate disposal system. This has several advantages including high-temperature incineration with long residence times and the incorporation of any ash into the cement product, thus avoiding disposal costs.

Other thermo-chemical treatments applicable to high-hazard wastes include pyrolysis and wet air oxidation. The latter of these involves heating the waste in a water slurry at high temperatures and pressures in the presence of air or pure oxygen (O_2). The products of this process are similar to those generated by combustion.

Biological methods of high-hazard waste disposal have been used for some time. In one system, known as land farming, oily wastes are spread onto the soil. Decomposition may be enhanced by the addition of inorganic fertilisers and the periodic disturbance of the land using conventional agricultural implements. This generates the right conditions for the breakdown of the wastes by the naturally occurring soil microorganisms.

Wastes that are neither recycled nor destroyed must be disposed of. This can be done with much greater safety if the waste is immobilised first. The technologies used to do this involve either the incorporation of the waste into a solid matrix or its encapsulation within an impermeable polymeric cover. In addition to immobilisation these processes are variously referred to as stabilisation, solidification or fixation.

Solid matrices within which waste can be incorporated may be formed of cementitious or organic polymeric material. Alternatively, inorganic wastes may be turned into a glass (vitrified) or incorporated into ceramic artifacts such as bricks. Vitrification involves the formation of a melt at around 1300°C and is therefore highly expensive. Consequently, it is generally reserved for the treatment of highly hazardous materials and may become the preferred option for the treatment of highly radioactive wastes.

Historically, the bulk of high-hazard waste has been disposed of to landfill, often with little or no pretreatment. In 1985, in the UK about 2.75×10^9 kg a^{-1} of chemical waste was sent to landfill; this compares with a total of about 4.2×10^8 kg a^{-1} that was treated chemically, fixed or incinerated.

It is clear that ill-considered landfill practices have caused and continue to cause environmental damage at a large number of sites. Nonetheless, when carefully managed, landfill is still seen as a highly appropriate means of disposal for many highhazard wastes.

Modern secured landfill facilities are located in areas where groundwater contamination is unlikely. They are covered and lined with impermeable membranes and leachates are collected, monitored and treated. The site is divided into a number of areas called cells, into which wastes of known characteristics are placed. This avoids the codisposal of incompatible materials and facilitates future removal of waste for recycling or further treatment. Ideally, an extensive programme of air and groundwater monitoring should be undertaken prior to the establishment of the facility and during and after its operation.

Heightened public concern and the increased commercial pressure on land means that the continued use of landfill as the primary means of high-hazard waste management seems in doubt, at least in the more-developed countries. Programmes of waste minimisation and waste treatment are likely to become more prevalent.

Other procedures that are used for the disposal of high-hazard wastes include dumping at sea. This has been a matter of controversy for some time. In *relative* terms the amounts of waste disposed of by this means are small; nonetheless, the absolute quantities are significant. For example, in 1985 the UK disposed of about 2.3 x 10^8 kg a^{-1} of chemical wastes in this way. However, there have been political moves to curb this practice. It was agreed by the 13th Consultative Meeting of the London Dumping Convention that all sea-dumping of non-inert industrial waste should cease by 31 December 1995.

High-hazard wastes have also been en disposed of by placement at depth within the Earth, well out of the reach of potable aquifers. This has been done both within disused mines and by deep-well injection.

International Trade in High-hazard Wastes

In recent years, in the more-developed countries, there has been a progressive tightening of the legal frameworks that regulate the disposal of high-hazard wastes. In many cases this has increased the financial cost of disposal within the countries of origin, spawning an international trade in noxious waste. For example, legal exports of hazardous waste from Europe to less-developed countries total about 1.2 × 10^8 kg a^{-1}.

In some cases, waste is imported by countries that have appropriate facilities for its treatment and disposal. For example, in 1992, clinical waste was imported by the UK from Germany for incineration in a specialised facility near Heathrow airport. Unfortunately, there have been a number of instances where high-hazard waste has been exported to countries that do not have the necessary facilities to deal with it adequately.

Waste Minimisation, Cleaner Production and Integrated Waste Management

Historically, industrial waste producers have relied on the cheapest means of disposal. This frequently involved discharges of untreated noxious material into water bodies or dumping on unsecured landfill sites. It is now evident that such inappropriate waste disposal practices have left a legacy of problems. The consequent economic, environmental and social costs are huge, but difficult to estimate. Most of these costs have been absorbed by society as a whole. However, there is evidence that attitudes are changing as the 'polluter pays' principle becomes more widely established.

Appropriate treatment and disposal of wastes can greatly ameliorate their environmental impact. However, this 'end-of-pipe' technology cannot reduce the amount of waste generated. This can only be achieved by in-process modifications targeted at cleaner production. The concept of cleaner production involves the application of integrated strategies aimed at avoiding unnecessary waste production and ensuring that the remaining wastes produced are innocuous. In order to be fully effective, this concept must be applied throughout the life-cycle of a product, from the extraction of the raw materials from which it is made through to its ultimate disposal.

Large manufacturing organisations are under increasing legislative and consumer pressure to limit the impact of their operations on the environment. There are now many examples of corporate initiatives that are aimed both at waste minimisation through cleaner production and at appropriate waste treatment. These can have direct economic as well as environmental benefits. For example, the 3M Corporation introduced its Pollution Prevention Pays, or '3P', programme in 1975. This concentrates on waste reuse and the reduction of pollution at source. The corporation believes that between its inception and 1989 the 3P programme directly resulted in a saving of US$408 million. Other examples include Polaroid's TUWR (Toxic Use and Waste Reduction) programme, started in 1987, and Dow's WRAP (Waste Reduction Always Pays) policy initiated the year before; excellent accounts of these measures are given by Buchholz.

Since the mid-1970s national and international agencies have been instrumental in promoting responsible waste reduction, treatment and disposal. The then EEC took an early interest, organising in 1976 one of the first meetings to discuss 'Low and NonWaste Technologies' (LNWT). By 1978 compendia of these technologies were available. Other initiatives include PRISMA (Project on Industrial Successes with Waste Prevention) in the Netherlands, the Environmental Management Company (CETESB) in Brazil, and UNEP's International Cleaner Production Information Clearing House (IPIC). For further information about these and other schemes.

Clearly there is still enormous scope for improvement. In many cases this can be achieved by simple means, such as waste segregation. To cite but one example, in Britain each hospital bed generates about 18 kg per week of waste. This includes everything from used dressings to flowers, all of which is classified as clinical and incinerated as if it were hazardous. In Germany, by careful segregation of truly hazardous materials from the rest, the amount of clinical waste is reduced to about 18 kg per year per bed.

8

Industrial Waste and Treatment

Industrial and municipal solid wastes are created by modern societies as unavoidable by-products of mining, industrial production, and the requirements of today's modern consumers. In the United States, the 19th century was an era of rapid growth during which many new and developing industries were established to systematically process raw materials into finished goods. During the 19th and 20th centuries, and how the 21st century, these industries concentrated on the quality of the final product, while discarding the residues and waste products generated during increasingly complicated manufacturing processes.

At the beginning of the Industrial Revolution, mining of coal and metal ores became both the fuel and the building blocks of Western society's industrialization. With the discovery of petroleum and natural gas deposits, chemical-processing industries quickly followed. From the early 1900s to the late 1960s, hundreds of thousands of petroleum-derived and synthetic chemicals were used in the production of goods.

Although carbon-based plastic materials, together with such organic chemicals as pesticides and solvents, have dominated industrial production since the early 1950s, metal-based goods remain fundamental to modern industry. Numerous modern goods—from cars to paints—require the use of common metals such as iron, aluminum, and copper. In addition, less abundant but more toxic metals like lead, cadmium, nickel, mercury, arsenic, and selenium are also used by industry. Metallic elements are therefore commonly found in industrial wastes, where they have complex and incompletely understood effects on the environment. Consumers also discard massive amounts of hazardous wastes. The advent of the development of countries such as China and India has exacerbated the production of industrial wastes. We buy, use, and dispose of increasing quantities of goods that have hazardous characteristics including paints, solvents, pesticides, batteries, lightbulbs and electronic goods.

The disposal of wastes can result in adverse effects not only on the environment, but also on human health. Early in the 20th century, industries, cities, and towns began to use abandoned or natural depressions to dispose of the evergrowing amounts of municipal waste residues. In essence thee were the earliest forms of landfills. Until the 1970s in the U.S., it was common practice to dump waste organics on land with little or no regard for the soil and water pollution it produced. Today, there are an increasing number of waste minimization programs recycle and reuse programs, and numerous solid waste disposal programs, which not only regulate hazardous industrial wastes, but also provide guidelines for the safe recycling and disposal of hazardous household wastes.

RELEVANT REGULATIONS FOR INDUSTRIAL AND MUNICIPAL SOLID WASTES

Before 1970s, industrial untreated wastes were disposed of in unlined landfills or lagoons or discharged into surface waters. Other wastes were burned in open pits or in incinerators with no pollution controls. As a result, groundwater and surface water resources, such as lakes and rivers, were polluted. Besides the human toll associated with water and air pollution, animals and plants also experienced adverse effects due to industrial pollution associated with the accumulation of pollutants in the environment.

Since the late 1970s, several laws have been implemented to control the disposal of hazardous wastes to protect our health and the environment. Most of the materials classified as nonhazardous are regulated under such specific industry categories as mining and oilfield wastes. Of all the wastes generated in the United States, more than 95% are classified as nonhazardous municipal. Although industrial, mining, oil- and gasified wastes are regulated separately, in this chapter we will treat all of them as "*industrial wastes*."

Also since the late 1970s, industrial wastes classified as hazardous have received considerable attention due to their obvious potential for deleterious impacts on human health and the environment. Consequently, government scrutiny has focused on new regulations for storage and disposal of these wastes. Nonetheless, less regulated, nonhazardous wastes constitute the bulk of the waste generated by industry. The largest source of nonhazardous wastes is human sewage, which is treated within wastewater treatment plants prior to land application, incineration, or landfilling. Nonhazardous wastes often contain the same potentially polluting components found in hazardous wastes, but at much lower concentrations. To prevent the disposal of household goods with hazardous properties into *municipal solid waste* (MSW) landfills, new guidelines have been implemented that control the disposal of items such as spend crankcase oil, care batteries, tires, paint thinner, or refrigerants. These are implemented through household hazardous waste collection programs.

Major Forms of Industrial Wastes

Regardless of how wastes are technically categorized by regulatory agencies, their pollutant-releasing capacities depend on their physical characteristics, which determine their mode of transport into the environment. The four major waste types based on physical characteristics—combustible wastes, solid wastes, sludge and slurry wastes, and wastewaters, which shows how each of these types can eventually release pollutants into the atmospheric, terrestrial, and aquatic environments. *Combustible wastes* yield by-products that can be released directly into the atmosphere as gases or particulate pollutants when they are not properly filtered. *Solid wastes* can release pollutants into the atmosphere via dust or particular transport, or when these wastes come into contact with water, their soluble constituents can be leached out into the soil surface or below. *Sludge* and *slurry wastes* can release pollutants into the soil and groundwater from both their solid and liquid phases. *Wastewaters*, owing to their liquid state, are always potential sources of pollution of discharged directly into the aquatic or terrestrial environment without proper treatment. In general, we can see that the pollutants whose impact on the environment is most severe are most usually found in liquid or gas phases. By their fluid nature, these pollutants can be transported large distances and thus affect large segments of the environment.

Except for a few isotopic forms, none of the 40 elements that are economically important to modern industrial society can be made synthetically. Thus, these elements have to be mined, extracted from the natural state, and subsequently purified. Because of such processes, as well as the processes used in manufacturing, these elements—including many precious and strategic metals such as gold (Au), platinum (Pt), cobalt (Co), antimony (Sb), and tungsten (W)—can be found in significant amounts in some industrial wastes. A few elements, such as mercury (Hg) and arsenic (As), can be found in gaseous, liquid, or solid wastes. Yet others, such as lead (Pb) and chromium (Cr), can be discharged into the atmosphere as particulate matter associated with other elements such as sulfur (S) and carbon (C).

However, most of the elements listed in Table 25.1, including metals, metalloids, and salts can be dissolved in the liquids (usually the aqueous phase) found in wastewaters, sludges, and solid industrial wastes. That is, with few exceptions, most industrial liquid, slurry and sludge wastes have a water phase that can range from 99% to less than 10% by weight. Therefore, the process of dewatering can be used to reduce much of the mass of these wastes. Moreover, many forms of organic pollutants are also water soluble to various degrees, and are therefore found in the water phase of these wastes These aqueous phases, once separated from solids, must be processed as wastewaters prior to discharge into the open environment.

Treatment and Disposal of Industrial Wastes

Technologies and practices used for the treatment and disposal of metal- and salt-containing wastes very widely, but ultimately, the by-products of these technologies end up being released into the air, land, or water environments. Methods that separate metals form the other waste constituents are driven by the need to reduce waste disposal costs and ultimately by the potential costs associated with liability. Organic wastes that contain low residual concentrations of metals and salts can often be degraded by using thermal or biological destruction processes that completely transform the waste into carbon dioxide and water. Conversely, wastes that contain significant amounts of metals and salts always leave indestructible residues that may or may not be recycled economically. Therefore, these wastes that cannot be eliminated, and must be disposed of in a manner that minimizes their impact on the environment.

Gas and Particulate Emissions

Gases and dust particulates that generated in the thermal destruction of wastes and smelting can be prevented from escaping into the air using one or more of the following processes: electrostatic precipitators, baghouse and cyclone separators, and wet scrubbers. These technologies are expensive and difficult to operate efficiently. However, without this equipment, smelters and incinerators would discharge large quantities of toxic metals into the atmosphere, thereby contaminating large tracts of land. In the United States, the Clean Air Act requires the control of hazardous emissions by industries. Originally passed in 1970, by 1990 this act was extended to cover 189 industrial chemicals, requiring the installation of air pollution control equipment on all major industrial sources. These regulations also include requirements for the removal of metal-containing particulate matter (PM_{10}) from air emissions.

A very controversial issue is the reduction of trapping of greenhouse gases like carbon dioxide (CO_2) that are released by power-generating plants that use fossil fuels (coal, oil and natural gas). These gases are known to be linked to global warming. To date, the U.S. government does not have or participate in any program to control CO_2 emissions. While mandated in Europe and other countries around the world that agreed to the *Kyoto Treaty*, only a few U.S. industries have reduced CO_2 emissions. However, future options to aggressively reduce thee emissions will have to include federal- and/or state-imposed limits of emissions. In addition, trading programs to buy and sell CO_2 emissions may be implemented, and the expanded use of carbon sequestration techniques is likely. Technologies being evaluated include trapping and liquefying CO_2 gas from smokestacks, followed by storage by injection into depleted deep oil or geologic gas reservoirs.

Chemical Precipitation

Wastewater and slurried sludges can be treated with chemical agents to precipitate and remove metals from the rest of the waste components. For example, solutions containing alkaline materials such as calcium carbonate, sodium hydroxide, aluminum oxide, or sodium sulfide are commonly used to precipitate metals from waste streams. These chemicals help form insoluble metal hydroxides, carbonates, and sulfides. Similarly, aluminum and iron oxides absorb metals such as cadmium and metalloids such as arsenic, thus removing them from solution.

General precipitation reactions can be represented by the following:

$$M(free) + CaCO_3(slurry) \rightarrow MCO_3 + M(OH)_2 \text{ (precipitates)}$$

$$M(free) + Na_2S \ (pH > 8) \rightarrow MS \text{ (precipitates)}$$

where M = cd^{2+} or Zn^{2+} or Cu^{2+} or Pb^{2+}.

Flocculation, Coagulation, Dewatering-Filtration-Decanting-Drying

Particulate pollutants in wastewaters can be made to settle out quickly by using chemicals known as flocculants and coagulants. *Flocculants*, which react with dissolved chemicals, facilitate the formation of aggregates or clumps, which can be decanted or filtered out of solution. Conversely, *coagulants* destabilize colloids, thereby permitting suspended particles to form aggregates that can settle out of solution. These chemicals can be useful in removing all kinds of pollutants from wastewaters, including metals and organic constituents. Flocculants and coagulants include iron and copper sulfates and chlorides, as well as complex synthetic organic polymers.

Dewatering and drying is an acceptable option for waste reduction where the liquid components of a liquid waste, such as a slurry or wastewater, have very low levels of pollutants, and these can be e treated using conventional wastewater treatment methods or even discharged directly into the environment. In this case, liquids are separated form solids via several dewatering options that include air drying particle filtration, sedimentation, and decanting. The separating solids are then treated and disposed of as solid wastes. Pond used to store and dewater flue gas desulfurization waste, a by product of sulfur dioxide gas scrubbing generated during coal combustion.

Stabilization (Neutralization) and Solidification

Sludges and slurries containing metals must be chemically and physically stabilized prior to final disposal. Since metals are more soluble in water at low pH, acidic wastes must be neutralized with such basic compounds as calcite ($CaCO_3$) and hydrated lime ($Ca[OH]_2$) to form low water solubility metal-carbonates and metal-hydroxides complexes.

The process of waste solidification usually involve trapping or encapsulating the waste into a physically stable matrix. For example, when wet cement is mixed in with sludges, it forms a stable block after a few days of curing (drying). This solidification method encapsulates wastes in a matrix that is relatively low in porosity and cannot easily be deformed or cracked under typical landfill overburden pressures. Consequently, percolating water does not readily infiltrate into the matrix, and metals or salts are less likely to leach out. In some cases, waste products like fly ash, collected from electrostatic precipitators during coal burning, can also be uses to neutralize and encapsulate other wastes like metal waste streams, due to their strong cementing (pozzolanic) properties.

Oxidation

Carbon-based waste streams can be oxidized to detoxify and destroy organic pollutants These are two major types of treatment processes:

thermal and chemical. The overall goals are the same in both processes. Thermal oxidation reactions can be described as follows:

$$RCCNOCl + O_2 + heat \rightarrow CO_2 + H_2O + N_2 + NO_x + SO_x + HCl + \text{intense heat}$$

Thermal oxidation processes include incineration, which use conventional fuel-driven burners. Solid or liquid organic wastes having low water content but high heat values are good candidates for thermal oxidation because they burn hotly enough to sustain the energy these processes require. The major disadvantages of this process include high costs (usually more than $200 a barrel), limited reliability, and a negative public perception of their safety. In addition, large emissions of pollutants can result when the systems are not operated and maintained properly. Incinerators operate most efficiently when designs are tailored to specific waste stream characteristics. Thus, incinerators are not suitable for the efficient oxidations of mixed waste streams.

Chemical oxidation generally proceeds via the following reaction pathway:

$$RCCNOCl + (O_2, Cl_2, \text{ or } O_3) \rightarrow R'CO \text{ (detoxified)} + CO_2 + H_2O + N_x + Cl^- + \text{intense heat}$$

where R or R' is the remaining part of the organic molecule. For example, chemical oxidation is used in the destruction of cyanide-containing wastewaters. This process involves chlorination, which can be summarized as follows:

$$2CN^- \text{ (liquid)} + 3Cl_2 \text{ (gas)} + 6NaOH \rightarrow N_2 \text{ gas } (pH > 8\text{-}9) + 6Cl^- + 2HCO_3^- + 2H_2O$$

Many forms and combinations of chemical oxidation processes utilize chlorine (Cl_2), chlorine dioxide (ClO_2), ozone (O_3), or ultraviolet (UV) radiation to eliminate low concentration of organics from industrial wastewaters. The chemicals oxidized using these processes include acids, alcohols, and aldehydes (such as oxalic acids and phenols), as well as more stable chlorinated pesticides and some petroleum-derived solvents such as DDT or xylenes.

Landfilling

Industrial wastes that contain significant concentrations of metals that cannot be recycled or recovered economically are usually good candidates for landfill disposal, which is the disposal of waste materials within soil or the vadose zone. However, sludges high in metals must be neutralized and solidified prior to landfill disposal and only landfills approved for the disposal of solidified hazardous wastes may be used for the disposal of these wastes. Similarly, wastes high in salts may also be buried in special landfills or used as fill materials for road and dam construction. Although such wastes are not considered toxic, they are very soluble in water. Therefore, it is important to minimize water contact to prevent the leaching of salts into the environment.

Landfarming of Refinery Sludges and Oilfield Wastes

Landfarming techniques have been used to treat hazardous wastes, particularly sludges, whose major constituents are oil, sediments, and water. In the early 1980s, land treatment of hazardous oily wastes came under intense scrutiny by the U.S. Environmental Protection Agency (EPA). Land disposal restrictions, begun in early 1992, now prohibit land treatment of hazardous oily wastes. These restrictions have compelled the petroleum industry to look at alternative disposal methods, including landfilling and incineration. On the positive side, petroleum refineries have also been spurred to develop waste-minimization strategies. Oilfield wastes generated during well drilling activities are treated in treatment facilities that have to be permitted. These facilities dewater and lower the salinity and oil content of these wastes until they meet state-imposed criteria for their safe reuse as fill materials. In addition to residual amounts of oil (~ 1-5%), these wastes often contain significant amounts of metals such as zinc, and minerals like barite, and sodium chloride.

Table 8.1. Examples of water solubilities of minerals found in wastes

Mineral names and chemical composition	*Water solubility (mg L^{-1})*
Barite ($BaSO_4$)	2
Calcite ($CaCO_3$)	14
Gypsum ($CaSO_4 \bullet 2H_2O$)	2,400
Salt (NaCl)	370,000

Deep-Well Injection of Liquid Wastes

Deep-well injection of liquid wastes into the subsurface is another waste disposal method. This method greatly reduces the potential hazard posed by wastes through disposing of them in the deep subsurface. According to the U.S. EPA (2001), there are about 470 Class I wells located mostly in the oil-producing states in the central part of the United States and in heavily industrialized states such as California, Michigan, and New York. Examples of deep-well injection of liquid wastes include oilfield wastes (brines) metal-containing wastewaters and slurries and wastewaters with high concentrations of toxic organic chemicals such as chlorinated hydrocarbons, pesticides, and radioactive wastes. According to U.S. EPA (2001), most wells (350 out of 473) inject wastes classified as nonhazardous (mostly oilfield wastes (brines)).

Deep-well injection is most suitable for handling large volumes of liquid or slurry wastes that have a water-like consistency (low viscosity). Watery liquids are usually pumped down into confined, aquifer-like zones composed of highly water-permeable material, such as sandstone or limestone. Similarly, oily wastes can be deep-well injected into high permeability subsurface zones. The depth of the injection zones, usually hydrologically confined, ranges from about 200 to 4,500 m (600 to 13,000 feet), and most zones are located between 700 and 2500 m (2,00 and 7,000 feet).

Drilling and constructing wells for these depths can be very expensive; therefore deep-well injection is primarily used by the petroleum industry, which already possesses inservice oilfield wells. Once constructed, these systems are comparatively cheap to operate and maintain.

The practice of deep-well injection of wastes still poses some potential problems. Possible clogging of the injection zone due to solid particles or bacterial growth may occur, and the injected waste may contaminate resident groundwater. This possibility is of major concern if the groundwater is a current or potential potable water source. Waste injection can cause groundwater contamination in several ways: (1) direct injection into the aquifer used for drinking water; (2) leaking wells (waste leaks form the well bore into an aquifer used for drinking water); (3) movement of waste to a zone that supplies drinking water. Despite these concerns, a study by the U.S. EPA (2001) concluded that "Current Class I Well Regulations are protective of Human Health and the Environment."

Incineration

The process of incineration that destroys highly toxic and hazardous organic wastes differs from MSW incineration, where energy is often produced. In general, low-temperature (upto 850°C) and high-temperature (~1200°C) incinerations use energy to oxidize carbon-and water-containing wastes to CO_2, H_2O vapor, and HCl and NO_x gases. However, some incinerator can serve as heat-energy sources, especially when oily wastes, such as spent oils, are used as fuel. But these incinerators must meet stringent emissions levels. Incineration efficiencies are also closely regulated, and destruction of all organic compounds must exceed 99.99%.

Incineration cannot be used with wastes that have high concentrations of water and noncombustible solids, nor can it be used for radioactive materials. Moreover, incinerators are very expensive to build and maintain, and waste incineration expenses often exceed $500 a barrel. Thus, this technology is mostly limited to low-volume wastes such as medial wastes. Finally, besides the gases previously listed, incinerators emit small but significant amounts numerous toxic chemicals including dioxins, furans, polynuclear aromatic hydrocarbons, and metals such as, mercury, beryllium, and cadmium. Often incinerators produce ash residues (bottom and fly ash) that have hazardous characteristics (particularly fly ash) and must be buried in landfills approved to handle such wastes.

Thus, incinerators do not totally eliminate the problems of environmental contamination, and this has resulted in the technology being condemned by the public. Grass-roots organizations are currently fighting for the elimination of this technology as a waste treatment option, even as a renewable energy recovery source. Because of the public outcry and health-related concerns raised by the scientific community, incineration has undergone a large decline in the U.S. during the last 10 years. The number of *municipal solid waste* (MSW) incinerators and medical waste incinerator in the U.S. has decreased from a peak of 186 in 1990 to 112 in 2003. Also, the same source indicates that medical waste incinerators decreased from 6200 in 1988 to 115 in 2003.

Stockpiling, Tailings, and Muds

Mining activities produce vast quantities of tailings, which are usually stockpiled in the form of terraces on or near the ore-processing mills. The environmental impacts of mining activities, which vary widely form site to site, are usually associated with runoff or percolation of waters contaminated with sediments. They are also related to pH, metal solubility, salt concentrations, and the quantities of wind-blown particulates that are contaminated with metals. Oil-and gas-fields produce large quantities of well cuttings (muds). These well cuttings, which contain barite, salts, and crude oil, are usually stockpiled or treated until they can be used as fill materials. The potential for offsite release of pollutants from these sites is normally associated with water and wastewater releases that contain varying concentrations of these chemicals.

Reuse of Industrial Wastes

Because metal- and salt-containing wastes cannot be destroyed, when improperly disposed they can be a hazard to humans and the environment. At the same time, these wastes have a potential economic value that is becoming more evident as the quality of their natural sources diminish. Precious and strategic metals, such as gold, platinum, cobalt, antimony, and tungsten, are routinely mined out wastes that contain significant concentrations of these elements. However, wastes that contain less valuable metals, such as aluminum and iron, are far less likely to be treated for the removal of these elements. Even more improbable is the extraction of highly reactive metals, nonmetallic elements, and their soluble salts, such as magnesium and calcium sulfates and carbonates, which are found in most industrial wastes

Metals Recovery

Economically valuable elements such as precious metals can be recovered from waste streams by using complex chemical reactions, including chemical separation or precipitation. Silver, for example, can be recovered from photographic wastes by acidifying the liquid wastes and separating the Age sludge that precipitates. Subsequently, the supernatant can be neutralized and disposed of safely, while the Ag sludge is sent to a smelter for purification. Metals can also be made to react selectively with synthetic organic chemicals known as chelates and synthetic zeolites, which are porous aluminosilicate minerals. In such reactions, metals are grabbed or trapped by the chelates or sequestered in the internal structure of the zeolites, while other materials pass by. Once reacted, the metal of interest is either precipitated or extracted out of the waste solution or is removed with a sorbent. The recovered metal sludge can be smeltered and refined into pure solid metal. Metal-containing residues resulting from smelting and refining of ores can be further refined by using a combination of chemical and physical separation techniques and high-temperature furnaces.

Energy Recovery

The majority of wastes containing organic carbon forms have large quantities of stored energy. Thus, wastes containing high concentrations of reduced carbon (usually organic carbon), such as organic liquids, woody materials, oils, resins, or asphalts, can be used in incinerators and electrical generators as energy sources. For example, kerosene and natural gas have about 44×10^6 to 49×10^6 Joules of energy Kg^{-1}.

Many solvents such as hexane, xylenes, and paraffins, which are found in oils and alcohols, have similar stored energies (42×10^6 to 58×10^6 Joules Kg^{-1}). Therefore, many waste mixtures of organic chemicals have energy values approaching those of commercial fuels. However, the limitations of this technology, when applied to wastes, are regulated by very strict air emission standards promulgated under RCRA.

Industrial Waste Solvents

Industrial wastes high in solvents are being successfully recycled using recovery systems that include distillation techniques and chemical and physical fractionation processes. For example, spent solvents used to clean and paint metal parts can be redistilled as a means to separate the heavy impurities form the volatile solvents. Some examples of solvents that can then be reused via solvent recovery stills include chloroform, acetone, benzene, xylene, hexane, and methylene chloride.

Industrial Waste Reuse

Some mine tailings can be used as fills for earthworks. However, economic and potential liability issues related to transportation costs and potential releases of contaminants usually limit their reuse. Consequently, mine tailings are most often stockpiled in place. Oilfield wastes that are high in salts may be leached with water to remove the soluble salts. The salt-free muds are then dried and stockpiled for use as fill material. These treated solid oilfield wastes have been shown to be safe in earthworks and release no residual salts into the surrounding environment.

Coal-burning wastes such as fly ash and flue gas desulfurization sludges have been used successfully as soil amendments. Coal-burning flay ash has also been used as a fill material for earthworks. These materials can be good sources of gypsum and calcite, which can neutralize acidity in soils an replenish macronutrients such as Ca, Mg, and S, and even trace elements like Zn, Cu, Fe, and Mn. Fly ash additions to acidic agricultural soils have also been shown to improve the soil pit, porosity and overall structure.

Treatment and Disposal of Municipal Solid Waste

Municipal Solid Waste

Municipal solid waste (MSW) is the solid waste material commonly called "trash" or "garbage" that is generated by home owners and business. Most state and/or federal rules permit landfills to contain either hazardous or nonhazardous wastes, but more than 95% of landfilling involves MSW. According to the U.S. EPA (2003), municipal solid waste consists primarily of paper, yard waste, food scraps, and plastic.

In 2003, EPA estimated that the United States was producing more than 236 million metric tons of MSW each year, or about 2 kg of trash per person per day, or nearly 1.7 tomes more waste per capital than in 1960. Developed nations such as the United States and Canada lead the world in the production of trash, producing much more dischargeable material than developing countries. More encouraging in the fact that trend sin MSW recycling have been increasing steeply since the 80s, now exceeding 30% in the U.S.

Although it was hoped that much of the landfilled material would biodegrade, in reality, most of the land filled waste from the past 40+ years is still present at landfill sites. Surveys of landfills have shown that conditions are often not favorable for biodegradation. Such conditions include low moisture, low oxygen concentration, and high heterogeneity of materials, many of which are nondegradable or very slow to degrade. Thus, old landfills serve as "*chemical repositories*," releasing pollutants to the groundwater and the atmosphere.

The number of landfills in the United States has been deceasing, from 18,500 in 1979 to less than 8,000 in 1988 and 1,762 in 2002, due mostly to the closure of small landfills. However, it appears that the dire predictions of landfill shortages by the turn of the century have not materialized. This is due in part to increase in recycling and the ongoing construction of much larger modern landfills. This is despite new regulations and landfill designs that have increased construction and permitting costs substantially.

Modern Sanitary Landfills

Until the late 1970s, waste materials were buried or dumped haphazardly within soil or the vadose zone, then simply covered with a layer of soil as a cap. There was no regulation of these landfills nor were protective mechanisms in place of prevent or minimize releases of contaminants from the landfills. Old landfills were usually located in old quarries, mines, natural depressions, or excavated holes in abandoned land. Conditions in these landfills are conducive to pollution, as these sites are often hydrologically connected to surface streams or groundwater sources. Water pollution is caused by the formation and movement of *leachate*, which is produced by the infiltration of water through the waste material. As the water moves through the waste, it dissolves components of the waste; thus landfill leachate consists of water containing dissolved chemicals such as salts, heavy metals, and often synthetic organic compounds. In addition to contaminating water, landfills release pollutants into the air. Anaerobic microbial processes generate greenhouse gases, such as nitrous oxide (N_2O), methane (CH_4), and carbon dioxide (CO_2).

A modern *sanitary landfill* is designed to meet exacting standards with respect to containment of all materials, including leachates and gases. New design specifications are predicated on minimal impact to the environment, both short-and long-term, with particular emphasis on groundwater protection. Landfill site selection is based on geology and soil type, together with such groundwater considerations as depth of water and use (i.e., aquifer vulnerability). In this design, the new landfill is located in an excavated depression, and fresh garbage is covered daily with a layer of soil. The bottom of the landfill is lined with low-permeability liners made out of high-density plastic or clay. In addition, provisions are made to collect and analyze leachate and gases that emanate from the MSW.

New landfills can cost up to $1 million per hectare to construct. Additionally, they require costly permanent monitoring for potential contaminant releases to the surrounding environment. Thus, many communities are faced with difficult decisions with respect to disposal of MWS–despite the fact that landfill design has improved efficiency with respect to pollution prevention, and increased recycling rates have slowed down MSW production.

Perhaps the most difficult problem associated with construction of new landfills is that of locating or sitting a landfill. Few communities want landfills located nearby. Real or perceived local community concerns about potential health effects and property devaluation often delay the location of a landfill for months or years. Thus, the costs of MSW disposal will continue to rise as new landfills are located in increasingly remote areas. However, increasing costs may force communities to look for alternative forms of MSW disposal, such as the development of new, more cost-effective recycling and waste-minimization strategies. Currently, in the United States, 30% of MSW is recovered, recycled or composted; 14% is burned at combustion facilities; and the remaining 56% is disposed of in landfills.

Reduction of MSW

The primary methods used to reduce MSW are recycling or composting (30%) and combustion (14%). These approaches are widely used in densely populated areas where land scarcity limits the use of landfills. In Europe and Japan, for example, less than 15% of MSW is sent to landfills, in contrast to the U.S., where over 55% is sent to landfills.

Combustion or incineration allows the heat derived from burning to be converted to other forms of energy. *Incineration* usually involves combustion of unprocessed solid waste; however, in some instances nearly 25% of the waste is processed into pellets, termed *refuse-derived fuel*, prior to burning. A new technique called *mass burning* can burn MSW at temperatures up to 1130°C, trapping the resultant heat to generate steam and electricity. Incineration is a relatively efficient method of reducing MWS, often reducing it as much as 90% by volume and 75% by weight. However, this technique is not without drawbacks. The primary problem associated with incineration is air pollution, due to the release into the air of toxic chemicals like dioxin and metals like mercury. This has led to a growing public resistance to this technology. Additionally, incinerator ash contains concentrated toxic metals and refractory organics that often require their disposal as hazardous waste.

In contrast to incineration, *source reduction* is aimed at preventions; thus, it is fundamental way of reducing MSW, because it tends to eliminate the need for landfills. Source reduction can be accomplished by a variety of methods, including the use of less material for packaging and the practice of municipal composting. For example, plastic containers may be reduced in size or eliminated altogether, and yard waste may be separated from other sources of MSW and used as compost. Municipalities often sell the composted material for use as a soil amendment or fertilizer.

Recycling, whose trends and growth rate have been previously discussed, involves the collection of certain types of trash, separating them into their components, and subsequently using those components to make new products. Materials that can be recycled include aluminum cans, plastics, glass, paper, cardboard, and metal. Also, as previously indicated, MSW may contain significant amounts of hazardous waste, which to date is not regulated under PCRA. However, in many communities, household hazardous waste programs have been implemented to recycle these wastes and prevent them from entering MSW landfills. Among the hazardous wastes or wastes with hazardous properties that have been banned and thus nearly eliminated from landfills include lead car batteries, tires, and spent motor oil.

Municipal sewage is treated within wastewater treatment plants and over 60% of the biosolids that are produced form treatments, are land applied. The remainder of biosolids are either incinerated or landfilled.

Pollution Prevention

Industrialization nations like the U.S. and many European countries are controlling sources and releases of pollution into the environment through implementation of the concept of *pollution prevention*. Pollution can be reduced or in some cases eliminated either by reducing the demand for products or by improving the treatment and manufacturing of raw materials and finished goods. Demand-driven pollution can be mitigated by the discovery and use of alternate but equally acceptable less-polluting products. Source reduction implies the modification of chemical processes or the development of alternate processes that ultimately produce less waste or less toxic waste.

In recent years, development of pollution prevention strategies has been driven by stricter environmental regulation and economic factors. For example, new industrial processes that emphasize waste reduction are not only economically desirable, but also increasingly required by government regulations. Thus, pollution prevention strategies are usually process driven.

9

ENVIRONMENT AND REPRODUCTION

The reproductive success of organisms depends on the interaction between their genetic endowment and the environment. As species evolve, those populations or individuals whose young are born at the optimal time for growth, development, and survival have the best chance to pass their genes to the next generation. Optimal time is used here to describe the period with the best combination of such conditions as availability of food, suitable temperature, and low incidence of predation.

Some vertebrate species, especially among fish (e.g., salmons) and birds (e.g., geese, the swallows of Capistrano) have used their rather efficient means of travel to migrate to favorable breeding grounds and to spend either their growing period and/or their nonreproductive period in environments that are most favorable for those parts of their life cycle.

For migratory as well as for nonmigratory species, natural selection has favoured the retention of those control systems that use the most reliable information predicting the environment and that cause reproduction to occur at an optimal time for survival of the young. Baker (1938), in a classical paper, proposed two sets of causes for determining the optimal time of reproduction, i.e., (1) ultimate causes, which exert selection pressure so that *populations* that reproduce during the optimal season survive; and (2) proximate causes, which provide *individuals* with the appropriate information, so that reproductive processes are initiated and continued in such a way that the young are born at the optimal time of survival. A priori, it appears that for aquatic animals, temperature or, depending on the depth at which they live, temperature and light would provide information that reasonably predicts future conditions. For most terrestrial organisms, day length is the most reliable source of information. We shall discuss the roles of various components of the environment but especially of light.

Endogenous Rhythms

In order to understand how environmental information is integrated, or in other words, how environmental stimuli exert their effect, it is necessary to consider endogenous rhythms first, because endogenous and exogenous rhythms interact with each other to stimulate or terminate reproduction. The two principal endogenous rhythms with which we will be concerned are the circadian and the circannual rhythms.

Light can have two functions in setting the stage for the reproductive season: (1) it can serve as a signal that entrains an endogenous biological rhythm, or (2) it may serve as a stimulus of the neuroendocrine system, such that gametogenesis, spermation, ovulation, endocrine stimulation of secondary sex organs and sex characters, and behaviour occur, thus permitting the animal to breed. An example of light functioning as a signal is the 6-hour shift of locomotor activity of house sparrows (*Passer domesticus*) when the lights on-off schedule is changed from lights on at 600, lights off at 1800 to lights on at noon, lights off at midnight. An example of light serving as a stimulus in the increase in testicular size when photosensitive white-crowned sparrows (*Zonotrichia leucophrys gambelii*) are transferred from a short day (8L:16D) to a long day of 16L:8D.

Circadian Rhythm

The role played by a circadian rhythm of photosensitivity in the photoperiodic response of organisms has been studied for a wide range of organisms and responses, such as flowering of plants, diapause in insects, emergence of pupae of *Drosophila*, and the testicular response of birds and fishes. It lies outside the scope of this book to consider the important theoretical aspects of this aspect of circadian rhythms of photosensitivity.

Two models have been proposed to explain the photoperiodic response of the neuroendocrine system that affects avian gonadal growth (which has been studied mainly in males): the external coincidence and the internal coincidence model. In the external coincidence model, it is assumed that light acts as a signal to synchronize the endogenous, circadian rhythm of photosensitivity, and that light, acting as a stimulus, falls in a period of photosensitivity and thus induces the response. This model is supported by two types of experiments: (1) resonance experiments, in which light-dark cycles are used with a relatively short light period, e.g., 8 h in "days" of 24, 36, 48, 60, 72, 84 h; and (2) by the photoscan or interrupted-night type of experiments. After a light period of 6 or 8 h (to signal the onset of dawn), the subsequent dark period is interrupted by a short (15 min-2 h) light period that comes at different times during the night for different experimental groups.

In the internal coincidence model, light serves as a signal and not as a stimulus, and of one can replace light by another signal, such as temperature change, one should obtain the same type of response to light as to a temperature change. In the internal coincidence model, it is assumed that there are at least two oscillators in which one is phased by down or by "lights on" and the other by dusk or by "light off." When the phases of these two internal oscillators are in an optimal configuration, the maximal response, e.g., testicular growth, occurs. This model is supported by two lines of evidence. In sone line of evidence, light as a signal can be replaced by a change in temperature; for example, a temperature cycle should phase the two oscillators, and the same type of response should then be found after exposure to a temperature cycle as after exposure to a similarly timed light cycle. Apparently, there has not been an experiment in which this has been tested for the avian gonadal response.

In another line of evidence, the organism may be exposed to a stimulatory light cycle for a relatively short period, so that the full response is not yet obtained, and the organism is then transferred either to continuous darkness or to a short photoperiod, e.g., 8L:16D. If the internal coincidence model is applicable, then the response should continue under DD, i.e., the testes continue to enlarge; whereas under 8L:16D, the response should diminish, i.e., the testes should regress. Vaugien and Vaugien (1961), using English sparrows and Wolfson (1966), using white throated sparrows (*Zonotriehia albicollis*), have indeed found that after transfer from long days to DD, the testicular growth was sustained and after transfer to short days, the testes regressed. Farner *et. al.* (1977) have pointed out that the number of birds used in each experiment by Vaugien and Vaugien (1966) and by Wolfson (1966) was low and that lack of measurement at critical time points in the experiment leave the data open to other interpretations. From their own experiments, in which they investigated the applicability of the internal coincidence model on English sparrows, Farner *et. al.* (1977) concluded that the internal coincidence model is too complex for the explanation of their findings. They concluded that their results could be explained on the basis of an external coincidence model that included a carryover phenomenon of the neuroendocrine events; i.e., gonadotrophins were released tonically after the stimulatory photoperiod was ended, thus sustaining gonadal growth Rutledge (1980) investigated whether the internal coincidence model applied to the starling (*Sturnus vulgaris*).

Male starlings were placed under (1) DD, after involution of their testes, following a natural photoperiod, or (2) a natural photoperiod followed by LL, or (3) under 12L:12D 45, 90, or 135 d after the testes had recrudesced and involuted. The time it took to obtain spermatogenesis was significantly shorter after exposure to natural photoperiods than after exposure to any of the other light regimes (about 60 d versus 100-130 d). These results suggest that in this species, an internal coincidence model may indeed be applicable, because it seems unlikely that a carry-over effect could explain the similar results obtained for the groups that were held 45, 90, and 135 d on 12L:12D, after testicular involution under the 12L:12D light regime.

The testicular response to variations on photoperiod has been investigated in most detail in birds. Farner *et. al* (1977) state that in more than 50 species in 15 avian families, it has been demonstrated that day length serves as important environmental information in the control of reproduction. In some species for which measurements are available, the logarithmic rate of testicular growth is linearly correlated with day length over the range of 9-10 to about 20 h. This has been found, e.g., for the house finch (*Carpodacus mexicanus*), the white-crowned sparrow (*Zonotrichia leucophrys gambelii* and *Z.leucophrys pugestensis*, the slate-coloured junco (*Junco hyemalis*) the green finch, (*Carduelis chloris*) and the Japanese quail (*Coturnis coturnix japonica*). The English or house sparrow is an exception this general relationship between day length and rate of testicular growth. In this species, in addition to the correlation between day length of above 9 h and testicular growth, there is also a testicular growth rate that is an inverse function of day length of less than 7 h. For the other species named, it appears that the external coincidence model alone can explain the observation, whereas in the English sparrow, it may involve an external coincidence model plus a "carry-over" phenomenon.

Circannual Cycle

An alternative to the idea that testicular growth is the result of light acting as either an entrainer and a stimulus (external coincidence model) or as an entrainer for circadian oscillators (internal coincidence model) is the idea that light acts as an entrainer for a circannual gonadal cycle. In order to discuss the possibility of such light-entrained circannual cycles, it is necessary to review briefly what the evidence is for circannual cycles. It is clearly, a priori, more difficult to establish the existence of an endogenous circannual rhythm than of a circadian rhythm. Moreover, there are important questions as to the definition of constant conditions, as Farner and Wingfield (1978) have pointed out. Does it mean that conditions are the same from day to day (e.g., 12L:12D throughout the year) or does it mean that conditions are constant each day and thus throughout the year? Very few experiments have been carried out under LL or DD to study gonadal cycles, except for the study in Paris, in which Pekin drakes (*Anas platyrhynchos*) were kept under LL for 54 moths and under DD for 70 months, and controls were kept under natural photoperiods. The drakes under LL had such a high mortality that the experiment could not be extended beyond 54 months. Spectral analysis of the data for the darkes under DD showed that there were two periodicities, one with a duration of 156 d and one with a duration of 319 d. Drakes kept under natural photoperiods had corresponding periods with durations of about 140-147 d and 365 d. Assenmacher (1974) proposed three possible explanations for the testicular cycle in the darke: (1) a circannual clock, and light acting as the Zeitgeber; (2) a circa-semestrial clock either entrained by light or with incomplete synchronization and a biphasic profile of testicular activity; (3) two endogenous rhythms fluctuating at a circa-semestrial clock either entrained by light or with incomplete synchronization and a biphasic profile of testicular activity; (3) two endogenous rhythms fluctuating at a circa-semestrial and a circannual periodicity. Assenmacher (1974) suggests that the second explanation may explain the data better than the other two.

Gwinner (1975), in a summary of evidence for avian circannual rhythms, lists circannual testicular cycles for the garden warbler (*Sylvia borin*) the blackcap (*S. artricapilla*), the African weaver finch (*Quelea quelea*) and the starling. It should be noted that of the species named, only starlings were kept under LL, whereas the other species were kept under which birds are kept is an important factor in revealing the existence of a circannual rhythm. For instance, startlings exposed to 11 h or fewer of light or to 12 h or more of light do not show a circannual testicular periodicity, but starlings kept under 12L:12D show such a cycle.

An alternative to the rhythmic variations in photosensitivity for gonadal stimulation is, that there is the possibility of an additive effect of photoperiod on the photoreceptor-hypothalamo-AP system. This effect would be comparable to an hourglass, whereas the models based on a circadian rhythm would be comparable to an oscillator. There appears to be only one example of such an hourglass model for gonadal stimulation in vertebrates, i.e., the American chameleon (*Anolis carolinensis*).

Influence of Light on the Neuroendocrine-Gonadal Axis

The onset of the reproductive season in many North Temperate Zone birds is timed by day length. The end of the reproductive period in several species is the result of the onset of refractoriness of the neuroendocrine-gonadal system. The gonads regress in spite of the fact that the photoperiod that initiated the onset of reproduction continues. The "site" of the refractoriness is not established, and it may vary among species. We will discuss briefly the evidence for each of the components of the neuroendocrine-gonadal system that is considered responsible for photorefractoriness.

There is apparently no inherent mechanism in the testes that causes regression. Starlings and canaries (*Serinus canarius*) maintained under 11L:13D show continuous spermatogenesis for extended periods of time. Injections of gonadotrophins in photorefractory birds will induce resumption of spermatogenesis. These types of experiments thus argue against the idea that testes become refractory. The photorefractoriness might, however, be the result of a negative feedback effect by gonadal hormones. By determining the LH concentration in the plasma of white-crowned sparrows, Mottocks *et. al* (1976) were able to show that photorefractoriness occurred in intact and in castrated birds. It thus appears that a negative feedback by gonadal hormones does not play a crucial role in photorefractoriness.

It is conceivable that the hypothesis becomes refractory to releasing-hormone stimulation; however, the experiments by Schwab (1971) cited above seem to argue against the idea that it is the inherent pituitary failure to respond to long term stimulation that causes the photorefractory state. Farner and Wingfield (1978) suggest that the site of photorefractoriness is either at the hypothalamus or at a higher level in the brain.

Experiments with starlings and domestic drakes (*Anas platyrhynchos*) have show that the thyroid influences plasma LH concentrations and the onset of testicular regression at the end of the breeding season. Wieselthier and van Tienhoven (1972) found that thyroidectomy of starlings prior to exposure to 17L:7D resulted in a failure of testicular regression at the time of regression in sham-operated controls. When thyroidectomy was carried out 4 weeks after the onset of exposure to 17L:7D, testicular regression followed, but after this a second period of testicular growth started, suggesting that thyroidectomy stimulated gonadotrophin secretion and the effect of thyroidectomy was not the result of an increased testicular sensitivity to gonadotrophins. The thyroid was not required for regression itself, and thyroidectomy prior to the refractory period did not affect the time of onset of the refractory period. The failure of refractoriness to occur after thyroidectomy before exposure to long photoperiod, is probably an indirect result of the high gonadotrophin secretion in this species after thyroid removal.

Jallageas and Assenmacher (1979) reported that domestic darkes, thyroidectomized in April, showed an increase in plasma LH concentrations in May; whereas in controls, plasma LH concentrations started to decline. In the thyroidectomized drakes, plasma LH concentrations reached a peak in August, then declined sharply; in the controls, plasma LH levels reached a nadir in July, then rose to a small peak in August, which was followed by a decline. Plasma testosterone concentrations were higher in thyroidectomized than in control darkes, and the postnuptial decrease in plasma testosterone concentration was delayed from June in the controls to October in the thyroidectomized birds. It is of particular interest that in intact drakes, plasma thyroxine (T_4) concentrations show a sharp peak in June-July. After castration, plasma T_4 concentrations were higher than in intact drakes, but in May they declined gradually and in July-August steeply, to reach control concentrations in February. Jallageas and Assenmacher (1979) speculate that there is "a temporary escape of thyroxine (TSH) secretion from the inhibitory effects of elevated testosterone levels, which would lead to an annual thyroxine peak in June and to the resulting gonadotropic depression."

It thus seems that in the starling and drake, the thyroid influences the onset of regression, but that this hormone does not affect the refractory period itself. Nevertheless, thyroid hormone affects the expression of the photorefractory state either temporarily, as in drakes, or permanently as in Indian finches, i.e., the weaver bird (*Ploceus philippinus*), the black-headed munia (*Munia malacca malacca*), the spotted munia (*Lonchura putnctulata*), and the chestnut-bellied munia (*M. atricapilla*). In these finches, the gonads remain at their maximal size for an indefinite period aftér thyroidectomy.

The photorefractory and photosensitive phases of the reproductive cycle of the white-throated sparrow may be controlled by different phase relationships between prolactin and corticosterone secretion. As Farner and Wingfield (1978) have pointed out, whether this mechanism applies to other avian species has not been tested.

It should be kept in mind that in species that breed early in the season in the temperate zone, young birds will be exposed to long photoperiods, which in adults cause gonadal recrudescence. Such young are photorefractory, but it is not known what mechanisms are involved in causing this refractoriness. It may be comparable to the onset of puberty in mammals, but not enough is known to make such a statement more than speculative.

The photorefractory state that is usually maintained during exposure to "long" days can be dissipated by exposure to "short" days. An extensive treatment of the physiology of the refractory period can be found in Murton and Westwood (1977).

Water Temperature and Day Length

In the teleost fish, the three-spined stickleback (*Gasterosteus aculeatus*), certain combinations of water temperature and day length interact with internal factors in regulating the breeding season. These interaction arc also important during the juvenile photorefractory period, which, of course, prevents reproduction in the young at an early age. Long days and high water temperatures do not stimulate the gonads when they are in phase 0, the condition found in immature young. However, short days and high water temperatures, such as prevail in late summer and early fall, can cause gonadal development to phase 1a. Gonads in this phase of development respond to the short days and low temperatures that prevail in the winter, and gonadal development proceeds similarly in adults and juveniles.

Juvenile stickelbacks exposed to 16L:8D at 20⁰C from July failed to show sexual maturation during a year to exposure, whereas sticklebacks kept on 16L:8D from July and exposed to 20⁰C water from July until August, to 4⁰C for 24 days and then to gradually increasing water temperature (from 4⁰C to 20⁰C in 18 days) matured in December (Baggerman, 1957). Thus, in this species, the temperature of the water is important in dissipating the juvenile refractory period when the fish are exposed to 16L:8D.

At the end of the breeding season (which lasts until about July in the Netherlands), the gonads regress. Under laboratory conditions, male three spined sticklebacks kept under 16L:8D and 20⁰C showed repeated alternate breeding and non breeding cycles, with a successive decrease in the duration of the nonbreeding cycles from an average of 108 d for the first nonbreeding period to 46.7 d for the second and 25 d for the third nonbreeding period. Thus there seems to be a refractory period, but unlike conditions for the birds discussed, there is no requirement for exposure to short days to dissipate the photorefractory period.

The testicular response of adult stickelbacks to 16L:8D to 20^0C in the fall is not substantially affected by prior exposure for 45 days to 16L:8D at 6^0C, 8L:16D at 20^0C or 8L:16D at 6^0C. However, both the stage of development of the gonads and the prior light and temperature regimes affect the response to 8L:16D at 20^0C. None of the fish exposed first to 8L:16D at 20^0C responded, thus demonstrating a refractory condition. Three of four fish with testes initially in phase 1a and exposed first to 8L:16D at 6^0C responded, but only one of four with testes in phase 1b responded to 8L:16D at 20^0C. Thus, the refractoriness of the testes to a change in temperature from 6^0C to 20^0C depends on the stage of testicular development.

None of the males with testes in phase 1a and exposed first to 16L:8D at 6^0C responded to 8L:16D at 20^0C, whereas six of eight males initially in phase 1b responded. Therefore, in this species, interactions between temperature, photoperiod, and the internal status of the animal prior to exposure to a stimulatory combination of temperature and photoperiod are important.

Environmental Stimuli and Reproduction

Photoperiod and temperature or a combination of them are probably the principal proximate causes that supply information to animals of the temperate zones for prediction of the optimal period for mating.

For animals that live and breed in the tropics, one would expect factors other than daylength and temperature to time the breeding season, if there is one. Similarly, for opportunistic breeders, such as the budgerigar, other factors would be expected to be important. Moreover, the exact timing of breeding of animals in the temperate zones might be expected to be affected directly or indirectly by factors other than photoperiod and temperature.

We will discuss, by given a few examples, the particular factors that are important, and in selected cases, the pathway that are involved.

Cyclostomes

Temperature of the water is the principal factor that times the migration upstream of *Lampetra flauviatilis* and *Petromyzon marinus*. Temperature is also the main factor for the timing of spawning. Hardisty and Potter (1971) list the various species of lampreys and the temperatures at which they spawn.

Spawning and courtship behaviour, in contrast to migratory behaviour, occur during the day and apparently with a preference for lighted areas. During courtship, the female passes the posterior part of the body near the male's head, suggesting that olfactory stimuli may play a role in the final stages of mating and spawning.

Elasmobranchs

In elasmobranchs, the main proximate cause is probably the temperature of the water.

Teleosts

For nontropical teleost fishes, photoperiod and temperature are probably the main proximate causes in timing the reproductive season. In the three-spined stickleback, there is evidence that the threshold value for gonadal stimulation by photoperiod varies throughout the year, it being high in June-July and low in March-April. In the rainbow trout (*Salmo gairdneri*) which spawns in the fall, gonadotrophin secretion and spermatogenesis are advanced by exposure to decreasing photoperiods and high temperature is similar to that in the male. Most freshwater teleost fishes spawn in the spring, and usually a combination of increasing photoperiods or of long photoperiods in combination with an optimal water temperature, which may be different for different species, bring about an increase in gonadotrophin secretion and recrudescence of the gonads. For the goldfish (*Carassius auratus*) the optimal temperature range for gonadal growth is between 17^{0}C and 24^{0}C. At 30^{0}C gonadotrophin secretion is stimulated, but the testes remain small and ovarian follicles become atretic.

The Pineal and Photostimulation of Reproduction

The way the photoperiod has its effect on gonadal recrudescence may vary according to the species. Several lines of evidence suggest that the pineal is involved in the photoperiodic response of some teleost fishes. In the eastern brook trout (*Salvelinus fontinalis*), the pineal organ is involved in (1) the entrainment of circadian activity by light-dark cycles, (2) the diurnal variation of plasma and liver metabolites, and (3) hypothalamic and endocrine cycles.

In certain fishes, e.g., rainbow trout (*Salmogairdneri irideus*) and in *Pterophyllum scalare,* the pineal has been demonstrated to have a photosensory function. In other species, there is morphological evidence for a secretory function of the pineal, e.g., the mosquito fish (*Gambusia affinis*) and *Symphodus melops*. In the golden shiner (*Notemigonus crysoleucas*), kept under 15,5L:8.5D, pinealectomy results in regression of the testes at 24°C and of the ovaries at 24° and 15°C. However, when this fish was kept under 9L:15D at 25°C, gonadal activity was increased by pinealectomy during the prespawning season (de Vlaming and Vodicnik, 1977). Pinealectomy caused disappearance of the daily cycle in GTH content of the AP, lowered the hypothalamic gonadotrophin-releasing hormone content of fish kept under 15.5L:15D at 25°C, and increased the hypothalamic GnRH content of fish kept under 9L:15D at 25°C. Thus the effect of pinealectomy on reproduction of this species depends on the photoperiod, water temperature, and the internal condition of the animal at the time of pinealectomy.

Pinealectomy of goldfish in either the fall or spring did not affect the gonadosomatic index, i.e., the gonadal weight per unit of body weight or the serum estradiol concentration. However, sectioning of the optic tract resulted in a lower serum estradiol concentration in the spring, but not in the fall. These results suggest that the effect of light, at least on estrogen secretion, may be mediated by a retinal pathway and that in the goldfish the pineal does not have a crucial influence on the secretion of estrogen. On the basis of the data available in the literature, de Vlaming and Olcese (1981) suggest that, although the results are not always consistent, the pineal may be a component in the pathway by which light affects the gonadal activity in goldfish.

In the Japanese rice fish (*Oryzias latipes*), the opposite response was obtained. In fish kept under LL, pinealectomy reduced the GSI in females, whereas blinding, which is presumably equivalent to DD, and affect the GSI, and blinded pinealectomized females had a similar GSI as pinealectomized females. When the GSI was measured regularly during the year, blinding was found to reduce the maximal GSI in controls kept under natural photoperiod almost one-half. Blinded pinealectomized females has an even smaller GSI than blinded ones. Unfortunately, no data for pinealectomized fish kept under similar conditions as the blinded and blinded pinealectomized ones were given concluded that the effect of light on reproduction of the Japanese rice fish is mediated through the eye and the pineal, with the slightly greater contribution made by the pineal, which he considers to be an endocrine as well as a photosensitive organ.

The pineal of fishes, as well as of other vertebrates, contains melatonin. Using the killifish (*Fundulus similis*), de Vlaming et. al (1974) have shown that melatonin injections cause reduction of the GSI in males and females kept under a long photoperiod in January. Male killifish, kept under long photoperiod (13L:11D) at the start of the spawning season (May), showed some reduction in the GSI after melatonin treatment. Males kept under a short photoperiod (19L:14D) showed almost a 25 percent increase in the GSI, a response that was reduced almost one-half by melatonin treatment. Thus melatonin must be considered a possible mediator in the photoperiodic response in this species. As we shall see in the discussion of mammals, in hamsters the role of melatonin can be either stimulatory or inhibitory, depending on the time of administration during the day, where as in the killifish, de Valaming *et. al*. (1974) found no difference in results following melatonin injections made at 2 h and 8 h after the onset of the photoperiod.

Other Influences

The reproductive cycle of the Indian catfish (*Heteropneustes fossilis*) consists of four periods: (1) a preparatory period (February-April), (2) a prespawning period (May-June), (3) a spawning period (July-August), and (4) a postspawning period (September-January). A threshold temperature of 25^{0}C is required for the formation of yolky follicles, and the rate of vitellogenesis is proportional to increasing temperature, once the threshold temperature has been reached. The preferred temperature, i.e., the temperature that the fish selects when given a choice, is between 28.6^{0}C and 32.0^{0}C. During the postspawning period, the follicles regress more slowly at 30^{0}C that at 25^{0}C, but eventually, regardless of photoperiod and temperature, the ovaries regress. Experiments in which this fish was kept under LL and under DD, indicate that there is an endogenous circannual cycle of ovarian follicular activity. It thus appears that daylight may act as a synchronizer of this endogenous rhythm and that water temperature regulates the exact timing of the reproductive cycle.

Visual stimuli play a role in determining the rate of reproduction and possibly the onset of breeding. By the use of colored models, it has been shown that visual signals are important in eliciting courtship behaviour in three-spined sticklebacks. Polder (1971) found that isolated *Aequidens portalegrensis* males and femâles showed no evidence of quivering, skimming, or spawning. However, all females spawned within four weeks when they were provided with a stone, on which to lay their eggs, and a mirror placed behind the stone in the aquarium. These females thus apparently were stimulated to reproduce by the visual stimuli of their own image. *Sarotherodon* (*Telapia*) *macrocephala,* kept in isolation so that they cannot see other members of their species, spawn less frequently than when they can see either a male, female, or gonadectomized conspecific in another aquarium. In these two examples, stimuli other than visual ones have thus been excluded. Similar observations have been made for *Sarotherodon mossambicus* was delayed by visual isolation, compared with young females that could see conspecific of the same age in another aquarium (Silverman, 1978b).

The European bitterling (*Rhodeus amarus*) shows courtship behaviour, nuptial colouration in the male, and ovipositor growth in the female only if certain fresh water mussels (e.g., *Unio*, *Anodonta*) are present. The female lays her eggs in such a mussel. In the absence of a male, she will still lay her eggs appropriately; therefore the interaction with a male is not required for spawning, although it is, of course, for fertilization of the eggs. We postulate here that visual stimuli are involved, but that other, e.g., olfactory stimuli cannot be excluded.

Olfactory signals apparently can be important in the courtship of the gobiid fish (*Bathygobius soporator*). Males of this species attack conspecific males and nongravid females that enter their territory, but gravid females are courted. Tavolga (1955) discovered that water in which gravid females had been present induced courtship behaviour in the male in a few seconds, but when the nostrils were plugged or cauterized this effect was not observed. The olfactory signal is apparently present in the ovarian fluid. In this same species, auditory signals also affect courtship behavior. The male can make grunting sounds, to which the female responds by darting movements. However, these darting movements show a definite orientation only in the presence of another goby, suggesting that visual and auditory stimuli are required for this behaviour to be meaningful.

Tidal effects apparently time the breeding of the grunion (*Leuresthes tenuis*) and the New Zealand white bait. The grunion's breeding season lasts from March to September. The fish land on the beaches during the high tide with one wave, and the males and females, with their genital openings in proximity, spawn in the sand, in which they have made a hole. With the next wave the fish swim back to the sea. The fertilized eggs develop in the warm wet sand and with the next high tide the embryos are returned to the sea, where further development occurs. However, although the spawning seems to be timed by the tides, the factors that time the maturation of oocytes and sperm have not been determined.

Tropical fishes do not experience a sufficient changes in the length of photoperiod to have photoperiod function as a signal for the timing of reproductive activity. However, the great differences in rainfall during different times of the year could be used as signals for timing of reproduction, either directly by changing the salinity of the water, or the water level, or indirectly, by increasing the food supply for the fish. For the tambaqui (*Colossoma bidens*), the jaraqui (*Prochilodus* sp), branquinha (*Anodus* sp), matrincha (*Brycon* sp), and peito de aco (*Patamorhina pristigaster*) of the Amazon basin, the rising water levels in the lakes in which these fishes live apparently regulates their time of breeding by increasing the amount of living space and the food supply. In general, one would expect availability of food, water temperature, and rainfall to be the most significant factors in the timing of reproduction of fishes in the tropics.

AMPHIBIANS

Anura

For anurans that breed in the temperate zone, temperature and rainfall appear to be the principal proximate causes. According to Jorgensen *et. al.* (1978), this effect of temperature may be partly an indirect one, in that it allows the animals (e.g., the toad, *Bufo bufo bufo*) to feed. This is an important consideration, since the nutritional status of the female is of the utmost importance in determining the growth of the follicles. In the fall and during hibernation, temperatures probably act directly on the animal, low temperatures preventing the follicles from becoming atretic. For anurans of the tropics, subtropics and deserts, rainfall, and thus probably also indirectly the availability of food may time the breeding period. For the edible frog (*Rana esculenta*), photoperiod as well as temperature are important in stimulating gonadotrophin secretion and testicular activity. Both the eye and the pineal, which in anurans has a photosensory function, are required for these responses to long photoperiods.

In courtship and the establishment of a territory, vocal communications are very important in many anurans, although there are exceptions, for example, the ascaphids, which do not have vocal signals. The volcalizations of the male green tree frog (*Hyla cinerea*) can be divided into three parts: (1) the so-called A calls, uttered when the male moves out to establish calling stations, (2) the B calls, which function in agonistic encounters and in the dispersion of the males, and (3) choruses of C calls that begin at about 2100 and last until about midnight. These choruses apparently attract the females, because by the time these choruses stop, all females found are in amplexus. Garton and Brandon (1975) speculate that in this species, it is the vocalizations rather than amplexus that induce ovulation. Oplinger (1966) has made a similar suggestion for the tree frog (*Hyla crucifer crucifer*), in which he found that ovulation preceded amplexus. He also suggested that humidity and vocalizations were the factors that timed ovulation.

The central neural efferent pathway for the control of calling in anurans involves (1) the ventral magnocellular preoptic (VMP) nucleus, which contains androgen receptors, although testosterone needs to be aromatized to estrogen in order to affect calling behaviour in the South African clawed frog (*Xenopus laevis*); (2) the ventral infundibulum, which like the VMP contains androgen receptors, and where testosterone is aromatized to estrogen in the South African clawed frog; (3) the dorsal tegumentum of the medulla, which concentrates dihydrotestosterone, and (4) the area of the hypoglossal and vagus nerve motor neurons, which also concentrate DHT in the South African clawed frog. Moreover, in this frog a nucleus of the sensory auditory pathway, the laminar nucleus of the torus semicircularis, concentrates DHT and estradiol, suggesting that testosterone is aromatized in this nucleus. These findings provide persuasive evidence that the androgen-induced calling is mediated by androgen affecting the efferent pathways at multiple sites and that the sensory pathway is also affected by the hormonal status of the recipient of the cells.

Narins and Capranica (1976) have investigated the neurophysiological aspect of the two-note "co-qui" call the neotropical frog (*Eleutherodactylus coqui*). In this species the resident male starts with a co-qui call. If he is approached by an intruding male, he drops the "qui" note and utters a "co" call. If the intruder approaches to within 0.6 m, he attacks the intruder. Females are attracted by either the two-note co-qui call or the "qui" call, but apparently not by the "co" call. Narins and Capranica (1976) found that this "dichotomy of the two notes in the male call reflects a difference in the distribution of the best excitatory frequencies of primary auditory neurons for males and females"; in other words, the male's and female's auditory system are "turned in" on different notes of the two-note calls.

We indicated that the release call by the female depends on the distension of the abdomen. Olfactory cues may have a function in attracting males to females. In the Surinam toad (*Pipa pipa*), the water in which a receptive female has been held, causes the male to become agitated and to attempt to clasp nonreceptive females.

Urodeles

In the temperate zone, temperature, humidity, and photoperiod are the most likely proximate causes for the breeding cycles of urodeles; in the tropics, rainfall is probably the principal factor. Olfactory signals are important in the courtship of the Plethodontidae, Desmognathidae, Salamandridae and Ambystomatidae, as discussed by Madison (1977). For the salamander, *Taricha rivularis,* there is evidence that odors secreted by the female can attract males, and the same may be true for *Triturus vulgaris*. Vocal communication is not pertinent in salamanders as they are avocal.

Reptiles

Tropical geckos and the tropical aquatic oviparous snake, *Hopolapsis buccatta,* have been reported to be continuous breeders. It is important to distinguish, however, between species in which breeding individuals can be found every month and between continuously breeding individuals, although, in each case there is no breeding season. Not all tropical reptiles lack breeding seasons; for example, the equatorial lizard (*Agama agama lionotus*) in Kenya lays eggs in July and August only. This cycle is timed by the availability of insects for food.

The proximate causes that time the breeding cycle has been reviewed by Lofts (1978), and the following summary is largely from his review article. Endogenous circannual rhythms may exist and time the reproductive cycle, but incontrovertible evidence for this is lacking. The single most important timing factor for most reptiles is temperature. For the Chinese cobra (*Naja naja*), Lofts (1978) speculates that there may be a thermorefractory period; i.e., gonads will not recrudesce after involution when the cobras are kept at the temperature that stimulates the recrudescence of the gonads. This is analogous to the photorefractory period in many avian species in which the gonads remain involved as long as the birds are kept under long photoperiods.

Day length affects the duration of the thermorefractory phase, with increasing photoperiods resulting in an increase in the duration of the photorefractory phase in the lizard, *Lacerta sicula*.

The American chamelon (*Anolis carolinensis*) has been studied in the greatest detail for the temperature-photoperiod interaction. The photosexual response, i.e., the increase in gonadal size in response to long photoperiods can be obtained in males at constant body temperatures between 28^0 and 32^{0}C, but not at 25^{0}C. When body temperature is not constant, the photosexual response can be obtained when the body temperature is 32^{0}C for about 8 h each day. As mentioned earlier in this chapter, the effect of photoperiod is additive (i.e., having hourglass clock).

Anolis carolinensis breeds in the spring when the neuroendocrine-gonadal system is not sensitive to photoperiod, and the activation of the gonads is the result of a response between July and October when day lengths are decreasing; however, when the animals are exposed to long days experimentally, the testes are maintained. In November, the sensitivity to photostimulation is lost, and temperature becomes the single most important signal for the timing of the reproductive phase. According to Licht (1972), the pathway by which temperature has this effect may involve: (1) a direct effect on the central nervous system, which then controls the secretion of gonadotrophin-releasing hormones, and (2) an effect on the response of the gonads to gonadotrophins. Licht (1972) has presented evidence for a strong effect of ambient temperature on the response of the gonads to exogenous GTH in the lizards, *A. carolinensis* and *Xantuxia vigillis*. For example, at 21^{0}C the ovarian weight of *A. carolinensis* after 12 daily injection of 19 mg ovine FSH was 9.7 + 0.4 mg; for the females kept at 29^{0}C, however, the ovarian weight after 10 daily injection of 10 μg ovine FSH was 161.7 ± 22.0 mg. There were no differences between the ovarian weights of control-injected animals kept at 21^{0}C and at 29^{0}C. Licht (1972) mentions two other possible pathways for the effect of temperature: (1) an effect on the AP, and (2) an effect on the half-life of the gonadotrophic and gonadal hormones, but neither of these possibilities was investigated experimentally. In this lizard, the pineal is apparently not required for ovarian recrudescence in the spring; however, in females kept on short days, 6L:18D at 31^{0}C, pinealectomy stimulates ovarian development, suggesting that on short days the pineal may secrete a substance that inhibits ovarian development either by acting on the hypothalamus, the AP, or the ovary. The substance may be melatonin, but the site of action has not been elucidated.

When a female is housed with one male, ovarian recrudescence occurs in nearly 100 percent of the females, provided that the photoperiod is 14L:10D, the temperature 32°C during the light and 23°C during the dark phase, the relative humidity 70-80 percent. When several pairs of these lizards are housed in one cage, the incidence of ovarian recrudescence drops to about 30 percent, if there is male-male aggression, but it is nearly 100 percent if the group is stable, that is, there is little male-male aggression. Females exposed to castrated males showed some delay in the growth of the follicles. It thus appears that courtship facilitates the ovarian response to a stimulatory environment and that male-male aggression inhibits this response. Long photoperiods have been shown to stimulate gonadal growth in the following species: MacCall's horned lizard (*Phrynosoma m' calli*), the Texas horned lizard (*P. cornutum*), the desert night lizard (*Xantusia vigilis*), the wall lizard (*Lacerta sicula*), the fringe-toed lizards (*Uma notata*, *U. inornata* and *U. scoparia*), and the red eared pond turtle (*Pseudemys scripta elegans*). In contrast to findings in the American chameleon, the response of the desert night lizard to long photoperiod is not affected by temperature between 8°C and 19°C. This differnece is explainable on the basis of ecological differences; however, the physiological basis for this difference has not been elucidated.

Thapliyal and Haldar (1979) have reported that in the Indian garden lizard (*Calotes versicolor*), pinealectomy early in the summer partially prevents the regression of the gonads that normally occurs upon exposure to 6L:18D. Pinealectomized males exposed to either natural daylight and temperature (10 h. 15 min light at 22 ± 2°C) or 15L:5D at 30°C, showed accelerated testicular growth and androgen secretion compared with the controls. This indicates that the pineal is inhibitory during long as well as during short days. On the basis of the information available, it is not possible to state what functions of the pineal of this lizard are essential for its effects on reproduction, photosensory function, and/or endocrine function.

Madison (1977), in an extensive review, states that in the lizard families, Iguanidae, Agamidae, and Chamaeleontidae, visual signals are the main releasers in courtship, whereas in the Scincidae, Lacertidae, and Teiidae, tectile, olfactory, and vocal signals are important. In the Igunidae, the visual signals are of primary importance, but olfactory cues are also used during courtship. In snakes, intraspecific communication relies mainly on visual and olfactory signals. For species and sex recognition, olfactory cues are of primary importance; visual signals play a minor role. Garstka and Crews (1981) found that yolk, or lipid extract of skin and serum of estrogen-treated female red-sided garter snakes when applied to the skin of males, elicited courtship behaviour by other males. The substance is probably produced in the liver under the influence of estrogen and may be associated with vitellogenin. According to Madison (1977), chemical cues are in all probability the main way of communication in turtles; however, visual cues cannot be excluded. The Crocodilia use visual, auditory, and chemical signals for courtship and territorial signaling.

Rainfall may time the breeding cycle of the females of several tropical species, e.g., *Anolis grahami*, *A. lineatopus*, and *A. sagrei.* However, the relationship between rainfall and ovarian activity seems to be based principally on correlations between rainfall and the onset of the breeding season, but no controlled experiments seem to have been carried out. Rainfall, of course, can act directly or indirectly by increasing the food supply.

Birds

Some of the breeding cycles encountered among birds are: (1) continuous breeding, (2) breeding intervals of less than a year, (3) six-month cycles, (4) opportunistic breeders, and (5) yearly cycles. We also mentioned in this chapter that there may be endogenous circannual testicular cycles in birds that breed in the temperate zone and that changes in day length may synchronize such cycles to yearly cycles. We will now discuss some of the environmental factors that time the onset and cessation of the breeding period.

Photoperiod

Photoperiod is the factor most investigated, especially for birds that breed in the North Temperate Zone. Generalization that can be made about species for which photoperiod is the main proximate factor are:

1. The rate of gametogenesis in photosensitive individuals within a species is determined by the length of the photoperiod; thus, the start of the breeding period is timed precisely.
2. The length of the photoperiod determines the increase in testicular size, probably because it determines the level of LH secreted in response to long days.
3. The duration of the response or the latency between the onset of exposure to long days and gonadal regression is determined by

day length; the longer the day length, the shorter the latency. Storey and Nicholls (1976) have speculated that the longer days stimulate a higher level of secretory activity by the neuroendocrine system and that this system may become exhausted sooner than when the days are shorter, but still stimulatory.

4. The duration of the refractory period is proportionally related to the day length; the shorter the day length, the shorter the duration of the photorefractory period.

Storey and Nicholls (1976) have suggested that the duration of the photosexual response may be directly related to the duration of the period of pretreatment with nonstimulatory day lengths.

All but the first generalization are illustrated by observations made on English sparrows: In populations of these sparrow at 34^0N (maximum day length about 15 h), the testes reach a size of about 400 mm^3, spermatogenesis lasts about 138 d, and the refractory period lasts about 13 d. Populations at 52^0N have a testicular size of about 500 mm^3, spermatogenesis lasts 106 d, and the refractory period is about 100 d.

Among closely related species, those species exposed to the longer day lengths have a larger ratio of testes to body weight than species exposed to shorter day lengths. Within a species, a more northern population may thus start to breed later because the day length threshold is reached later than for a more southern population; however, the more northern population would show more rapid testicular growth than the southern population. This rapid growth is an adaptive response, because the duration of a favorable period to breed and raise young is shorter for the northern population than for the southern one. This variation in testicular growth with latitude has in fact been found for chaffinches (*Fringilla coelebs*) in Russia.

Transequatorial migrants encounter special problems, because they will experience long days during their stay on the wintering grounds. Apparently the bobolink (*Dolichonyx oryzivorus*), which is a transequatorial migrant, has a slow response to 14--15-h light periods, and its refractory period can be dissipated by 12-h photoperiods, a photoperiod not effective in dissipating the refractory period of white-throated sparrows and juncos (*Junco hyemalis*).

Changes in day length are too small in tropical species to play a role in timing of the breeding cycle; nevertheless, he African weaver finch (*Quelea quelea*) can respond to exposure to different photoperiods, although the response is not the same as in most species in the North Temperature Zone. *Quelea quelea* kept under 12L:12D had enlarged testes for about 7 months, then had a 42-d refractory period, and subsequently showed testicular recrudescence. When the birds were transferred from 12L:12D to 17L:7D, the testes initially increased in size, but the testes remained enlarged for a shorter period than those of birds kept on 12L:12D. When the birds were transferred from 17L:7D to 8L:16D, the testes remained regressed. After the birds were transferred from three weeks on 8L:16D to 17L:7D, the testes showed recrudescence, but at the same time as birds that had been kept on 17L:7D throughout. This and other evidence indicate that the photorefractory period in this species is not shortened by exposure to short days. Exposure of *Quelea quelea* to the natural photoperiod to which English sparrow are exposed in the North Temperate Zone resulted in a gonadal cycle quite similar to that of English sparrow exposed to the same day lengths. Conversely, English sparrow subjected to weekly stimulatory photoperiods has gonadal cycles that resembled those of *Quelea quelea.* Readers interested in other types of breeding cycles are referred to Murton and Westwood (1977), who discuss many different types of breeding cycles more extensively than can be done here.

Mention should be made of the lack of an inhibitory effect of continuous darkness (DD) on reproduction of Australian grass parakeets or budgerigars (*Melopsittacus undulatus*). Females provided with a nest box and exposed to male vocalizations start laying after a latency that is the same for birds exposed to 6L:18D, 2L:22D, or DD, but that is longer than for birds on 14L:10D. In the absence of a nest box, however, this latency is shorter for birds in DD than for birds under 14L:10D. Apparently the DD to some extent simulates the darkness of the next box, and such darkness may be a requirement for the female to reproduce. When given skeleton photoperiods consisting of an initial 6L followed by 2L after various periods of darkness, 6L:6D:2L:10D proved to be the most effective skeleton photoperiod of this basic pattern for temperate zone birds.

The pathway by which photoperiod induces gonadal recrudescence has been extensively investigated in darkes, Japanese quail, English sparrows, and white-crowned sparrows. The evidence demonstrates that blinded birds can show this photosexual response, and so suggests the presence of an extra-retinal receptor, as proposed originally by Benoit in 1935. Further investigations with different techniques by which light was directed at exactly localized places in the brain have revealed that extra-retinal receptors are located in the hypothalamus. The most precise localization has been accomplished by Yokoyama *et. al* (1978) in the white-crowned sparrow. These investigators found that receptors for the photosexual response were located in the ventromedial hypothalamus and/or in the tuberal complex, and perhaps also in or near the nucleus rotundus in the thalamus. Oliver (1979), using Japanese quail, found that phosphorescent pellets of 0.6 x 0.2 mm induced the photosexual response when implanted in the infundibular region, but not when placed in either the preoptic region or in any region of the hypothalamus.

The fact that there is no anatomical evidence for presence of photoreceptors or photopigment in these photosensitive regions of the hypothalamus makes the interpretation of the mediation of the effect of light difficult. A thermal effect does not seem to be likely, but it cannot be excluded. However, the fact that infrared stimulation of sites that respond to visible radiation does not provoke a photosexual response argues against a thermal effect.

The relative roles of the retinal and extra-retinal receptors in the photosexual response have been investigated by the use of blinded birds and birds in which access of light to the extra-retinal receptors was blocked, e.g., by the injection of India ink under the skin on top of the skull or by hooding of the birds. In English sparrow, white-crowned sparrows and golden-crowned sparrow (*Zonotrichia atricapilla*), blocking of the extra-retinal receptors decreased the testicular growth, migratory restlessness, and fat deposition, compared to controls. However, some increase in these three measurements occurred in the experimental birds after transfer from short to long days. These data suggest that with extra-retinal receptors blocked, the information about the photoperiod that is perceived by the eyes is ignored. There is, however, evidence that information through the retina may have an inhibitory effect. In Japanese quail, blinded males and females do not show gonadal regression when transferred from a long day to a short day, if the birds have experienced long days prior to blinding. In female white-crowned sparrows, information through the retina may inhibit secretion of gonadotrophic hormone (LH) and the final stages of development of ovarian follicles. This information may, of course, consists of either visual or photic stimuli.

On the basis of the evidence presently available it seems that the pineal is not involved in the photosexual response of either intact or blinded birds, although the pineal clearly plays a role in circadian rhythmicity of locomotor activity, body temperature, and migratory restlessness of several species, e.g., the house sparrow, white-crowned sparrow, white-throated sparrow, and house finch (*Carpodacus mexicanus*) . It is intriguing that in the adult chicken, the pineal seems to have no effect on circadian rhythmicity but that in the baby chick, the pineal in vitro retains a circadian rhythm of N-acetyl transferase activity and that such an explanted pineal is photosensitive.

Oliver (1979) has shown that in Japanese quail, hemispherectomy did not affect the photosexual response, that the area in which light could provoke this response was limited to the infundibular complex, and that red light (620 nm) was slightly more effective than green light 530 nm) in inducing the response. Electrolytic lesions of the posterior part of the basal infundibular nucleus or of the dorsal part of the infundibular complex (IC) or of the preoptic anterior hypothalamus block the photosexual response. Deafferentation of the basal medial hypothalamus prevents the photosexual response.

Oliver (1979) has explored, by electrophysiologic techniques, the extent to which the POA and the IC areas differ. He found: (1) shorter latency of evoked potentials in the POA than in the IC, (2) a greater reduction in spontaneous multiunit activity (MUA) in the POA than in the IC of quail on short days (8L:16D) by severance of the optic nerve, and (3) direct retinal connections to the POA, but not to the IC. In both areas, long days (18L:6D) diminished spontaneous MUA after light flashes. However, in the IC a decrease in MUA could be obtained either by exposure to 16L:8D or by implantation of radioluminous pellets, indicating that a retinal pathway to the IC was not required for this response to long days. Oliver (1979) suggests that there are two hypothalamic areas, one with a direct responsiveness to light, the other requiring retinal input. but both have the capacity to regulate gonadotrophin secretion.

Pinealectomy of Indian weaver birds (*Ploceus philippinus*) indicates that the pineal may have an antigonadal function; for example: (1) after pinealectomy of young birds, sexual maturity occurs precociously; (2) pinealectomy accelerates gonadal recrudescence in response to long photoperiods; and (3) the gonads recrudesce in pinealectomized birds exposed to short photoperiods (8L:16D) that normally do not stimulate gonadal growth. Pinealectomy has no apparent effect on the gonadal response to long photoperiods in the white-crowned sparrows (*Zonotrichia leucophrys gambelii*), Harris' sparrow (*Z. querula*), house finches (*Carpodacus mexicanus*), and Japanese quail (*Coturnix coturnix japonica*). Evidence that pinealectomy decreased the ovarian response of blinded mallard ducks, *Anas platyrhynchos,* to continuous light, and the testicular response of mallard drakes to natural long daylight in the spring is not convincing.

Photoperiod affects not only intact birds and gonadal activity, but it can also affect the response to exogenous hormones, as in the following examples:

1. Canaries on long days are treated with PMSG showed greater nest-building activity than PMSG-treated females under a short photoperiod.
2. The nest-building activity of estrogen-treated ovariectomized canaries, intact, photorefractory, estrogen-treated canaries, and castrated estrogen-treated males is considerably higher under 20L:4D than under 8L:16D. This difference between the results obtained under short and long days is, according to Hinde and Steel (1978), probably not the result of the greater amount of time available for nest building under long days than under short days.
3. A higher portion of estrogen-treated ovariectomized females enters the nest box under 14L:10D than under 8L:16D.

Temperature

Other factors than photoperiod influence the time of breeding. Temperature probably has a modifying effect on the photoperiodic gonadal response. Correlations between the onset of breeding and temperature may be the result of an indirect effect of the temperature on food supply, and availability of food is of particular importance for a ovarian growth because of the extensive metabolic demands made for yolk deposition. Murton and Westwood (1977) state, "In no temperate-zone photoperiodic species have any of these other factors been shown to cause gonadal recrudescence in the absence of appropriate photostimulation." Under experimental conditions, testicular growth of white-crowned sparrows (*Z. leucophrys gambelii*) is somewhat greater at higher temperatures (in the range of 5.2-34.1^{0}C) under long days

Visual Stimuli

The effects of visual stimuli have been investigated elegantly, both under laboratory conditions and under field conditions. In the rigdove (*Streptopelia risoria*), it was shown first that visual stimuli produced by castrated males were less effective in inducing ovarian and oviductal growth than stimuli from intact males. Subsequently it was demonstrated that ovarian and oviductal recrudescence was nearly linearly related to the dose of androgen administered to the males to which the females was exposed. The dose of androgen determined the courtship behaviour of the male, and the female's reproductive organ showed a corresponding response.

The visual stimulation provided by the female in turn affects the testicular activity of the male; however, this can be demonstrated only under short day length. Compensatory growth of the left testis in unilaterally castrated male ringdove was greater than in isolated males, but this occurred under short-day and not under a long-day regime, presumably because under the long-day regime the testis was already maximally stimulated.

A second example of visual stimuli on avian reproduction is found in the experiments by Smith (1966), in which the main features used by four sympatric gull species for "recognition" of their own species were the yellow eye ring and yellow iris in the herring gull (*Larus argentatus*),, the red-dish-purple eye ring and dark iris in Thayer's gull (*L. thayeri*), and the reddish-purple eye ring and irisies varying between very light and dark in Kumlien's gull (*L. glaucoides kumleni*). In these gulls, the female selects the male, and she chooses male with an eye-head contast similar to her own. In mated pairs, the female's eye-head contrast serves as a "releaser" for the male to mount. By capturing males and females and painting eye ring of different colour, Smith was able to deceive females into selecting males of a species other than their own. By subsequently changing the eye-head contrast of the female, so that it resembled that of the male's own species, the males were induced to mate with these females. When Smith changed the eye-head contrast of the female in male-female pairs of the same species, the testicular growth of the testes observed in controls was prevented. The pathways by which the visual stimuli affect the neuroendocrine-gonad unit need to be elucidated.

Olfactory Stimuli

In general, the olfactory system of birds is not well developed; however, Balthazart and Schoffeniels (1979) showed that after severance of the olfactory nerve in domesticated Rouen ducks (*Anas platyrhynchos*), sexual displays and social behaviour decreased. These investigators suggest that the differnece in chemical composition between secretions of the preen gland during the reproductive season and nonreproductive season may serve as a cue for the male. Sexual displays have an important role in courtship, and, as discussed for ringdoves above, such displays may affect the development of the females' reproductive system. It is thus conceivable that the olfactory system affects the interplay between male and female necessary not only for mating but also for development of the reproductive organs to their fullest potential.

Auditory Stimuli

Receiving auditory stimuli, or being able to give them, is important for full development of testicular or ovarian size in at least two species, the budgerigar (*Melopsittacus undulatus*) and the ringdove. Under experimental conditions, isolated pairs of budgerigars will not reproduce unless they can hear the vocalizations of another male, or even specific isolated parts of a male's vovalization, such as the tusk and soft warble (both of which are normally given during precopulatory, behavioural displays) when played on a tape recorder. The male of the isolated pair, however, needs himself to be able to vocalize in response to the vocalizations received from another male, in order to show full testicular development.

The time of the day that vocalizations are received affects the testicular response. Gosney and Hinde (1976) have shown that during a 14-h day the female's reproductive development is stimulated to a greater degree, if the male song is presented during the first half of the day than if it is presented during the second half of the day. Subsequently, Steel *et. al.* (1977) demonstrated that in ovariectomized birds, treated with estrogen and prolactin male vocalizations tends to increase the proportion of birds that entered the next box and the amount of time spent in the next box. They also showed that male vocalizations were more effective in the morning than in the afternoon of a 14-h day.

Although visual stimuli are important in inducing ovarian and oviductal growth in ringdoves, auditory stimuli also have a stimulatory function. Auditory stimuli form a breeding colony increased the response in a female exposed only to the visual signals given by a courting male. In the absence of visual stimuli by a courting male, auditory stimuli from a breeding colony could induce ovarian and oviductal growth, albeit not to the same extent as the visual stimuli given by a courting male. When vocalization of female ringdoves were altered, so that the cooing sound was changed by either severing the hypoglossal motor nerves to the syrinx or by lesioning the nucleus intercollicularis, the ovarian and oviductal response to a courting male were inhibited. Thus in the ringdove and in the budgerigar, the gonadal response to male stimuli requires either that the recipient be able to make or hear his or her own normal auditory responses, respectively.

In canaries under a short daylight regime (11L:13D) songs by the male accelerated ovarian growth and increased plasma LH concentrations, but this effect of auditory stimulation was not found under 14L:10D, presumably because the gonadal system was maximally stimulated by this photoperiod.

Rainfall

Rainfall is correlated with the reproduction of budgerigars, probably because after the rain, grass seeds can germinate and the grass will then form seeds, which then become a source of food for the birds. The reproduction of the African weaver bird (*Quelea quelea*) is also correlated with rainfall. These birds migrate from areas where the rainfall has started, which makes seeds unavailable because they start to germinate, to areas where rain has not started. After about a month, they return to the original area from which they migrated to harvest the seeds of the grasses that have bloomed after the rainfall. The availability of green grass to build nests probably is also important in timing reproduction. In both the African weaver bird and the budgerigar, neither the testes nor the ovaries regress completely between breeding periods, but they appear to be ready to develop as soon as the opportunity arises.

Captivity

Captivity can severely inhibit reproduction, especially among birds. In many species the male in captivity will show normal testicular development, but in females, ovarian development is not completed to the stage in which large yolky follicles are formed, as for example in the wild mallard duck (*Anas platyrhynchos*). In ducklings obtained from eggs collected from wild mallards, it has been possible to increase ovarian development by imprinting ducklings on the experimenter. Correlated with the increased ovarian development was a diminished fear response of the ducks when they were adult. The pathway by which captivity has its effect on fear behaviour and on the inhibition of ovarian development probably involves the archistriatum, which has projections to the hypothalamus. Lesions of the archistriatum diminished the fear response and increased ovarian development in wild ducks not imprinted on humans.

Nutritional Factors

The Pinon jay (*Gymnorhinus cyanocephalus*) is apparently dependent for its reproduction on the availability of green pine cones of *Pinus edulis*. Feeding on such pine cones in the fall induces testicular recrudescence in Pinon jays.

Subtle Factors

Brockway (1962) found that budgerigars did not visit the nest box and did not breed unless the vertical distance between the roosting perch and the nest hole was at least 10 cm. Within minutes after in adjustment was made to increase the distance from less than 10 cm to 10 cm, birds started to visit the nest box, and eventually breeding occurred. This requirement for a distance between a roosting perch and a nest hole may have evolved as a protection against predators.

MAMMALS

Photoperiod

Evidence that photoperiod is the proximate factor for the onset of breeding of mammals is limited to few species. Among mammals that respond reproductively to photoperiod, two groups can be distinguished: one group that breeds with the onset of long days and another group that breeds with the onset of short days. To the first group belong, among others, the golden hamster (*Mesocricetus auratus*), the voles *Microtus montanus* and *M. arvalis,* the Djungarian hamster (*Phodopus sungorus*), the raccoon (*Procyon lotor*), the ferret (*Mustela furo*), the mink (*M. vison*), the sable (*Martes zibellina*), the hare (*Lepus timidus*), and the horse (*Equus caballus*). To the second group, the short-day breeders, belong, among others, the sheep (*Ovis aries*), the goat (*Capra hircus*), the white-tailed deer (*Odocoileus virginianus*), and the silver fox (*Vulpes fulva*).

The pathway of the photoperiodic response involves: (1) the retina, (2) retino-hypothalamic and retino-hypothalamic-pineal connections, and (3) the pineal. The retina is required for the photoperiod to affect reproduction. Blinding of hamsters has the same effect on the gonadal regression of hamsters as exposure to short days. As discussed at the beginning of this chapter, the external coincidence hypothesis implies a time-measuring device in the organism, Turek and Campbell (1979) state, "the mammalian eye is clearly involved in the perception of light used to measure time," but they point out that the identity of the retinal photoreceptors is not unequivocally established.

The retino-hypothalamic tract, a tract from the retina which projects mainly to the contralateral suprachiasmatic nucleus (SCN), but with a few projections to the ipsilateral SCN, is required for the photoperiodic response. Experimental interruption of the retino-hypothalamic tract, while sparing the primary optic tract, also destroys the SCN because of its location. This poses a serious difficulty for interpretation of the data, because the SCN is probably the location of the biological clock, or it is such a crucial part of the biological clock, or it is such a curcial part of the biological clock that the clock cannot function after the SCN is destroyed. Rusak states, "No circadian rhythm has survived SCN ablation in rodents, but a variety of noncircadian cycles can be generated by lesioned animals." Thus lesion of the SCN interrupt not only the transmission of photic stimulation, but also the time-measuring mechanism.

After the SCN is destroyed, exposure to short days does not result in gonadal regression (Stetson and Watson-Whitmyre, 1976). Since blinding causes gonadal regression, it is clear that lesions of the SCN not only interrupt the transmission of photic stimuli to the hypothalamic-AP-gonadal system, but also have the effect of interrupting or destroying the circadian clock.

The pineal in mammals receives information about the photoperiod by a pathway from the retina→the retino-hypothalmic tract→the SCN→connections to the lateral hypothalamus→connections to the lateral hypothalamus→a connection probably through the medial forebrain bundle and the mid brain reticular formation to the upper thoracic intermediolateral cell column→preganglionic fibers to the superior cerivical ganglion→preganglionic fibers to the superior cervical ganglion→postganglionic fibers which form the nerve conarii→pineal.

The physiology of the photosexual response has been studied most extensively in the golden and the Djungarian hamster, the ferret, and sheep.

Golden hamster

The seasonal reproductive cycle of the golden hamster is ecologically explained as follows. In the fall, when day length becomes shorter, the testes regress, and as temperature drop, the animal will hide in a burrow and begin hibernation. During the hibernation in the dark, the gonads and secondary sex organs regress; presumably, this conserves energy. As spring approaches, the gonads recrudesce preparatory to the emergence for hibernation, so that when the weather becomes favorable, those animals that have fully functional gonads can start to reproduce immediately. The evolutionary advantage of such preparation is obvious.

The regression of the ovaries and uteri and of the testes and secondary sex organs that normally occurs in the fall can be induced by exposure of the animals to short days (1L:3D). In spite of continued exposure to short days, however, gonadal recrudescence takes place in about 20 weeks, a period equivalent to that spent by hamsters in burrows. After such gonadal recrudescence, the reproductive system is refractory to the effect of short days unless the animals are first exposed to about 11 weeks of long days.

Some physiological aspects of this refractoriness has been elucidated:

1. The amount of testosterone required to suppress gonadotrophin secretion in castrated golden hamsters is higher in hamsters under long days than under short days. This higher sensitivity decrease in hamsters kept on 6L:18D and this correlates with the dissipation of the refractoriness in intact control hamster.
2. Melatonin, a hormone secreted by the pineal, administered to hamsters with recrudescing testes under short photoperiods, does not inhibit the growth of the testes, indicating a lack of sensitivity of the larger organs for melatonin.
3. The induction of photorefractoriness is, according to Turek and Losee (1979), not the result of exposure to short days in themselves, but the result of an inhibition of the neuroendocrine-gonadal system. In melatonin-treated hamsters transferred to short days, the gonads did not regress as long as the melatonin was administered, but regressed after melatonin was no longer administered, indicating that the neuroendocrine-gonadal system still responded to short days. In hamsters receiving melatonin from week 2-24, testes were regressed at 12 weeks, active at 24-weeks, and active at 32 weeks, so that these animals were refractory, indicating that melatonin treatment did not prevent refractoriness.
4. Exposure to long photoperiods restores the involution response of the testes when hamster are subsequently exposed to short photoperiods. This dissipation of photorefractoriness is accompanied by a restoration of the sensitivity of the target organs to the inhibitory effect of melatonin. This restoration by long days of the inhibitory response to short days and to the inhibitory effect of melatonin on the reproductive system requires the presence of the pineal.
5. For photoperiodic stimulation of the testicular activity, the golden hamster has to receive at least 12.5 h of light per day if uninterrupted light-dark schedules are used.

The effect of pinealectomy on reproduction has been studied extensively in the golden hamster. The salient findings for the male are : (1) After blinding, by removal of the eye or after exposure to short photoperiods, such as 1L:23D, the testes and secondary sex organs regress and are atrophied after 10-12 weeks; however, after an additional 14-16 weeks, they recrudesce in spite of the blinding or the exposure to 1L:23D. (2) Either pinealectomy or superior cervical ganglionectomy (SCGX) prevents this response if carried out prior to blinding or exposure to 1L:23D, or reverses the regression if performed after the regression has started. (3) In general, FSH, LH, and prolactin concentrations in the plasma are lower in hamsters with regressed testes due to enucleation or exposure to 1L:23D than in comparably treated hamsters with active testes which have had a pinealectomy or a SCGX. (4) After implanting two pituitaries from donor hamsters under the renal capsule, in order to obtain high plasma PRL concentrations, and after twice-daily injections of LH-RH, the testes of hamsters enucleated or exposed to 1L:23D were similar in size to those of controls exposed to long photoperiods, whereas either transplanting two pituitaries alone or twice daily LH-RH injections alone failed to restore testicular size. (5) The inhibitory effect of the pineal on testicular size of golden hamsters exposed to 1L:23D is, in all probability, the result of melatonin secretion by the pineal. (6) The amount of testosterone required to suppress LH and FSH concentrations in plasma of castrated hamsters is considerably lower under short-day periods than under long-day periods. In other words, the neuroendocrine unit is more sensitive to the negative feedback by androgen under short-day than under long-day periods. This effect is at least partly mediated by the pineal.

In female golden hamsters, the following findings are important: (1) Enucleation or exposure to 1L:23D results, within several weeks, in acyclicity of the vaginal morphology as shown by vaginal smears, whereas in control females there is a rather rigid 4-day cyclicity. The ovaries become devoid of follicles but become heavier than those of controls because the interstitial tissue proliferates, and the uteri become atrophic in about 10-12 weeks. (2) Pinealectomy or SCGX prevents the effects of either blinding or exposure to 1L:23D, as it does in males. (3) The FSH and LH concentrations in hamsters blinded or exposed to 1L:23D are not different from control females under long photoperiods. The surges of FSH and LH which occur daily in the afternoon in controls, also occur in hamsters that are enucleated or exposed to 1L:23D. (4) The data on PRL concentration in the plasma are not sufficient to draw valid conclusion, although from the available data, it appears that plasma PRL concentrations are lower in the acyclic hamster than in the cyclic controls.

The identity of the pineal hormone that inhibits the gonads is not the evidence indicates that it probably is melatonin. According to Turek and Campbell (1979), melatonin is of most interest because: (1) the pineal is the principal source of melatonin; (2) synthesis and release of melatonin are influenced by photoperiod, with high secretion occuring during darkness; (3) melatonin, if injected into hamsters under a long photoperiod 6.5 h or later after light on, results in involution of the testes and cyclicity of the vagina shown in vaginal smears. However, when injected in the morning, melatonin has no such effect, and continuous administration of melatonin mixed in a pellet of beeswax implanted under the skin to provide continuous delivery of the melatonin resulted in maintenance of the gonads of male and female hamsters exposed to 1L:23D. This effect is exactly the opposite of the one expected, if melatonin secretion is to function as an inhibitor of the gonads. Reiter (1981) explains this finding as due to a decrease in meltaonin receptors as the result of continuously high melatonin concentrations.

The evidence presented by Reiter (1981) is persuasive for the experimental conditions under which melatonin is continuously supplied. It does not explain, however, why under 1L:23D, when there should be a continuously high melatonin secretion, such a decrease in melatonin receptors, and thus a stimulatory effect on the gonads or at least maintenance of the gonads, does not occur. Turek *et al.* (1976), using Silastic capsules to deliver melatonin continuously, found that melatonin suppressed testicular activity of golden hamsters kept under 14L:10D. Reiter (1981) states that he has been unable to repeat these experimental results. The reason for the difference between the results obtained by Rieter (1981) and by Turek *et. al* (1976) is not clear.

Turek and Campbell (1979) suggest that the following generalizations can be made about the function of melatonin in stimulating testicular growth of golden hamsters: (1) Melatonin is stimulatory for gonadal activity during exposure to short days, but is inhibitory to gonadal function when the animals are exposed to long days. (2) Melatonin inhibits testicular growth that is dependent upon long photoperiods, but it cannot inhibit testicular growth that occurs independent of long photoperiods, such as testicular growth during the first nine weeks of life or the spontaneous recrudescence that occurs when the golden hamsters are maintained under 1L:23D for about 28-30 weeks.

Arginine vasotocin (AVT) has been tentatively identified in the pineal of golden hamsters, but its effect on testicular regression has been mainly investigated in rats and mice; however, its effect on reproduction in the hamster is not determined. In bovine pineals, a polypeptide other than AVT, which also suppresses testicular activity of mice, has been isolated. It has been determined whether such a peptide is present in the pineals of hamsters and whether this bovine polypeptide affects the gonadal activity of hamsters.

The discussion of reproduction of the hamster to this point has been limited to the proximate causes which regulate the seasonal breeding. Within the breeding season, thee are communication systems between males and females. Vaginal secretions from diestrous and estrous females are attractive to males and these odors have sexual excitant effects. Singer *et. al.* (1976) isolated dimethyl disulfide from hamster vaginal secretions and found it to be an attractant for male hamsters.

In addition to olfactory signals, hamsters have vocal signals for communication. Female and male hamsters emit ultrasounds, which in each sex depend on the endocrine status of the female. The rate of ultrasound emission by the female is dependent on estrogen, and the emission by the male depends on signals emitted by the female. For example, either testestrone administration to ovariectomized females or hypophysectomy of adult females eliminates the emission of ultrasounds by the male. These auditory signals may function to advertise the presence of a female and her endocrine status to the male. The signals by the male also indicate the presence of a conspecific animal, and they have a facilitating effect on the female's reproductive behaviour, but the female does not give ultrasounds during lordosis. It thus appears probable that these signals are not important during heterosexual contact. The untterance of ultrasound signals by the male is androgen-dependent, for after castration the rate of ultrasound production is reduced, and androgen therapy restores the ability to produce ultrasounds at a rate similar to that of intact males.

Djungarian hamster

The Djungarian or hairy-footed hamster (*Phodopus sungorus*) lives in the steppes of Mongolia and western Siberia. This animal does not hibernate, but the changes in testicular weight and weight of the secondary accessory organ are similar to those described for golden hamsters. The size of the testes can be manipulated by changes in photoperiod. Exposure to short days cause regression of fully developed testes and exposure to long photoperiod causes a recrudescence of involuted testes. In contrast to golden hamsters, there are daily cycles of plasma testosterone and prolactin concentrations in the plasma, so that the time at which blood samples are taken becomes an important factor in making comparisons between Djungarian male hamsters with regressed testes on short photoperiods and males with large testes on long photoperiods. In Djungarian hamsters, pinealectomized males do not show regression of the testes upon exposure to short photoperiods, but in contrast to the findings in golden hamsters, pinealectomy with regression of the testes and secondary sex organs diminishes the increase in weight of the testes and secondary sex organs upon exposure to long photoperiods, i.e., photoperiods of 13 h or longer. In pinealectomized Djungarian hamster exposed to 8L:16D until 45 d of age, the testes do not reach the same size as the testes of either pinealectomized or intact males exposed to 16L:8D for the same length of time. This suggests that in the Djungarian hamster, not all effects of long and short photoperiod are mediated through the pineal. Melatonin administration to Djungarian hamsters has yielded contradictory findings, as it has in golden hamsters, and at present it is difficult to give a satisfactory interpretation of these results. Hoffmann (1981) concludes that the pineal in Djungarian hamsters is not specifically concerned with regulation of gonadal functions, but that it plays an integral role in the transduction of photoperiodic stimuli upon the neuroendocrine axis and thereby may influence many functions.

Ferret

The breeding season of the ferret (*Mustela putorius*) starts in the spring and lasts through August. Estrus starts in the spring and, unless ovulation is induced by coitus, lasts for 20-25 weeks. Exposure of estrous females to short photoperiods induces anestrus and exposure of anestrous females to long photoperiods induces estrus. However, animals kept on long days eventually end their estrus and then remain in anestrus, permanently, sas do several avian species, as long as they are receiving long photoperiods, suggesting that they have become photorefractory. After transfer to short days for 7 or 8 weeks, the animals are capable of responding again to long photoperiods by going into estrus. By exposing ferrets to 6-, 4- or 2-months cycles, of long and short photoperiods, 1, $1^1/_2$, and 3 estrous periods can be obtained in one year; however, alternating cycles of long and short days of 1 month fail to induce regular cycles, but yield, instead, irregular intervals between estrous periods. The length of the photoperiods does not affect the induction of estrus in ovariectomized ferrets treated with gonadal hormones, whereas in ovariectomized canaries, the nest-building activity induced by estrogen administration is greater under long than under short photoperiods.

When ferrets were blinded neonatally, they came into estrus in the spring, as in intact controls, but in subsequent years the controls came into estrus in the spring, whereas the blinded ferrets had recurring estrous cycles (for as long as five years), which were not synchronized to photoperiod or other factors. The fact that blinded and control ferrets came into estrus at the same time may be the result of a preprogrammed onset of puberty, which may be related either to age or to body weight. The recurrent breeding cycles of the blind animals being out of synchrony with the controls suggests that in the controls, light may act as a signal for synchronization of an endogenous circannual rhythm. If this were the case, then the blinded ferrets should show circannual rhythms of estrus. Herbert *et. al.* (1978) interpret their data as showing no evidence for such a circannual rhythm in the blinded ferrets. It appears, however, that no statistical analysis was carried out to ascertain whether there was a periodicity in the data. It would be worthwhile to analyze the data by appropriate statistical methods.

Pinealectomized ferrets resemble blinded ones in having recurrent unsynchronized breeding cycles, suggesting that both the pineal and the eye are required for the transmission of information about photoperiod to the neuroendocrine axis. The effect of the pineal on ferret reproduction does not seem to be mediated by melatonin. However, as discussed for the golden hamster, the time and method of administration of melatonin affect the response; therefore, no premature conclusions about the function of melatonin in the reproductive cycle of the ferret should be drawn. Herbert and Klinowska (1978) suggest that serotonin, which is present in high concentrations in the pineal of the ferret, and the SCN may be involved either in transmitting or modulating the effects of photostimulation by long photoperiods.

Mink

Under natural conditions, mink mate from the middle of January until the middle of March in the North Temperate Zone. The young are born near the end of May, independent of the time of mating because implantation does not occur until mid-March, when prolactin concentrations apparently increase sufficiently to induce implantation. Exposure of long photoperiods in late fall advances the time of estrus, and exposure to long photoperiods of mated females advances the time of implantation. A similar response of advanced implantation has been observed in the sable (*Martes zibellina*), the pine marten (*M. americana*), and he European marten (*M. martes*).

An unexpected effect of additional illumination (17L:7D) has been reported by Belyaev and Zhelezova (1978) for mink carrying the autosomal mutation shadow. This is a dominant mutation for coat color with a recessive lethal effect. All homozygotes die, mostly prior to implantation. By lengthening the photoperiod, Belyaev and Zhelezova (1978) were able to reduce embryonic mortality of shadow kits and to increase the ratio of shadow to nonshadow kits significantly by additional illumination. These authors also state that under long photoperiods, embryos heterozygous for shadow were less viable than normal nonmutants.

Belyaev et. al. (1975) have reported that in the Georgian white fox, there is an incompletely dominant autosomal coat mutation with a recessive lethal effect on the embryos, which die before implantation. As in the mink, the mortality was reduced by providing long days. Belyaev et. al. (1975) also found that as litter size increased, the survival of homozygous embryos increased. They ascribed the effect of long days and of large litter size on survival of these embryos to an increase of CL, activity; however, they gave no quantitative data on progresterone secretion. Long days did not favorably affect the survival of two other mutations, silver- and white-faced, in which homozygous embryos also die early.

Manipulation of the photoperiod is used in a few domesticated species either to improve reproduction or to obtain young at a specific time. In horse breeding, it is important to have the young born early in the calendar year because of the criteria used in defining 2-year-olds for competitive purposes. By using a lighting and temperature program that was phase shifted several months, Sharp and Ginther (1975) were able to induce estrus in 7 out of 7 mares by day 112 after the start of the increasing light and modified temperature program. Oxender et. al. (1977) found that exposure of mares to 16L:8D induced estrus and ovulation, which was accompanied by increases in LH and progesterone. In the control group, kept outdoors, estradiol peaks were higher than in the 16L:8D group, but there were not associated with estrus, whereas the smaller E_2 peaks in the 16L:8D group were associated with behavioural estrus; this suggests that factors other than E_2 concentrations are important in the induction of estrus in the horse. Supplemental lighting is used for mink and foxes to induce earlier breeding and, in the case of mink, to induce earlier implantation.

Sheep

The reproductive season of sheep in the têmperate zones starts in the late summer or early fall and seems to be initiated by decreasing photoperiods. By manipulation of the photoperiod, the onset of estrus can be induced at will; however, there is increasing evidence that there is an endrogenous rhythm. Under constant light regimes of 6L:18D, 12L:12D, 18L:6D or LL, first estrus occurred in ewes at about the same time as in ewes that had been exposed to a decreasing or to an increasing photoperiod. During $2^1/_2$ years on such constant photoperiods, all groups experienced three breeding seasons. This suggests the presence of an inherent rhythm, which controls estrus in the absence of changes in photoperiod but which will respond to changes in photoperiod.

Other evidence for an inherent circannual rhythm in sheep is found in the changes in plasma prolactin concentrations. In France, in ewes and rams under a natural photoperiod, plasma prolactin concentrations are high in June and low in December, and generally parallel those of the duration of daylight during the year. By the use of skeleton photoperiods, e.g., 7L:6D:1L:10D, it was demonstrated that there was a variation in photosensitivity, and that prolactin concentrations could be increased by maintained on natural daylight, 8L:16D, 7L:3D:1L:13D, 7L:6D:1L:10D, 7L:9D:1L:7D, or 7L:12D:1L:4D, the variations in prolactin concentration during the year were parallel for the different photoperiods.

After transfer of rams from 16L:8D, to 8L:16D, there is an initial rise in the plasma concentration of FSH and LH in about four weeks with peak concentrations at 5-9 weeks for FSH and 3-8 weeks for LH, followed by an increase in plasma testosterone concentration, peaking at 10-18 weeks and an increase in testicular diameter at 11-17 weeks, such a change in photoperiod also results in a decrease in plasma prolactin concentration. Lincoln and Peet (1977) also found that transfer from 16L:8D to 8L:16D increased the frequency of episodic gonadotrophin release (most LH peaks occur during the dark period). It thus appears that a combination of changes in the secretion of FSH and LH, which both increase in concentration and in frequency of episodic release, and prolactin, which decreases in concentrations in response to short photoperiod, stimulates testicular activity. The sudden transfer from short to long photoperiods resulted in progressive increase in plasma prolactin and a decrease in plasma FSH concentration.

The main difference between the breeding and nonbreeding season appears to be the effect of estradiol on LH secretion. When the negative feedback effect of estrogen is high during the nonbreeding season, LH is suppressed and estrogen does not have a stimulatory effect on LH secretion. This concept has been discussed in greater detail by Legan and Karsch (1979), and more supporting evidence is presented in their review paper, which should be consulted by the interested reader.

The effect of photoperiod on the reproductive system may be mediated through the hypothalamus, the AP, or the pineal or any combination of those. Investigations by Pelletier and Ortavant (1970) suggest that the LH-RH content of the hypothalamus of rams, as in hamsters, increases during photoinhibitory regimes (long days in sheep, short days in hamster), whereas FSH and LH concentrations decrease. This suggests that long days inhibit release of LH-RH. Land et. al. (1979) found that the response of ovariectomized ewes to LH-RH varied with the time of the year. This response would have to be investigated under carefully controlled photoperiods, before this variation can be ascribed to photoperiodic variation. Pinealectomy, either before or during the breeding season, did not affect the number of estrous cycles, the duration of estrus, the incidence of ovulations during anestrus, the LH concentration between peaks of the cycle, or LH concentration anestrus. In rams, pinealectomy resulted in a lower mean plasma prolactin concentration in the summer (14 ± 3 ng/ml versus 29 ± 2 ng/ml for controls), a higher concentration in the fall (53 ± 8 ng/ml versus 26 ± 3 ng/ml for controls), but no effect on spring and winter plasma prolactin concentrations. Control rams showed seasonal differences in LH peak frequencies, with higher frequencies in the summer, whereas pinealectomized rams did not show such a seasonal change. The peak plasma concentration of FSH found during the fall in controls was absent in pinealectomized rams, but there was no significant effect of pinealectomy on plasma testosterone concentrations during any of the seasons. After SCGX of rams in February, testicular recrudescence started and continued as in control rams. During exposure to sequential 16-week periods of short days (8L:16D) and long days, SCGX rams had a continuously high testicular weight, secreted high amounts of testosterone, FSH, and LH, and showed less variation in plasma prolactin concentrations between long and short days than controls. Control rams had smaller testes, lower plasma testosterone, FSH and LH concentrations, and higher prolactin concentrations, during long days than during short days. These data suggest that a functional pineal is not required for the normal seasonal change in testicular size, but that it is necessary for the response to abrupt changes in light regimes.

Goats

The endocrine profile of male goats under natural daylight conditions resembles that of rams, i.e., LH and testosterone start to rise in August, reach a peak in October, and decrease to basal concentrations in November to December. FSH shows a small, but not statistically significant peal in September and a large peak in April. Prolactin concentrations parallel the photoperiod. Exposure of female goats to 19L:5D, for 70 d starting at the end of January and subsequent exposure to natural photoperiods induced estrus, LH release, and ovulation 60-80 d in advance of control goats kept under the natural photoperiod.

Deer

The changes in the plasma concentrations of LH, testosterone, and prolactin in captive, male white-tailed deer, show a pattern similar to that found in sheep and goats; FSH concentrations are parallel with those of LH, and no peaks are found in the spring.

Swine

In wild pigs (*Sus scrofa*), reproduction is seasonal, with most estrous periods between December and February, although a sow many have another estrus in April or May if she loses her litter or if she has her young early and they are weaned before April. According to Mauget (1978), the reproductive efficiency of domesticated pigs in France decreases in the summer, the season in which wild pigs cease to reproduce. It appears, therefore, that the tendency for seasonal reproduction may still be present in domesticated pigs. In domesticated boars that were exposed to 15L:9D, puberty occurred earlier than in controls on natural daylight, suggesting that photoperiod may be one of the proximate cause in the seasonal reproduction of this species. Once puberty is reached, olfactory and auditory signals are important in reproduction of swine. It appears that the olfactory signals emitted by the male are of primary importance in bringing male and female together, whereas in sheep, male and female each is attracted by the signals emitted by the sexually active member of the opposite sex. The chemical signals emitted by the boar are metabolites of testosterone, i.e., 5 α-androst-16en-3-one and 3 α- hydroxy-5-androst-16-en-3on and 3 α- hydroxy-5-androst-16-ene. Testosterone is metabolized to these sex attractants in the salivary glands, and the material is excreted in the saliva. Females treated with testosterone became as attractive to an estrous female as to a boar. A sow or gilt in estrus will stand almost rigidly when touched on the lack. To obtain this so-called standing response, the sow has to be exposed both to olfactory signals, of which 5 α-androst-16-en-3-one is the most effective, and to auditory signals, such as the grunting by the boar. Additional visual and tactile stimuli add relatively little compared to the effect obtained by the olfactory and auditory stimuli. When artificial insemination is used commercially, it is sometimes difficult to determine which gilts or sows are in estrus and it is a common practice to spray 5 α-androst-16-en-3-one in the vicinity of the sow and then touch her back to determine whether she will stand rigidly and thus is in estrus.

Dietary Factors

In the mammalian species discussed so far in this chapter, photoperiod was the most important proximate cause in timing of the breeding season. In at least three species of voles. *Microtus agrestis*, *M. arvalis*, and *M. montanus,* which show seasonal reproduction, the timing of the breeding season is related to the availability of green feeds, such as grass, alfalfa, wheat. The growing season for these plants is, of course, in turn nodulated by photoperiod and temperature. In *M. montaius,* the reproductive activity seems to be triggered mainly through the availability seems to be triggered mainly through the availability of certain plant foods. Negus and Berger (1977) showed that supplying a natural population of this vole with fresh green wheat grass for two weeks resulted in a high incidence of pregnant females, whereas no pregnant females were detected in the control population. Berger et. al. (1981) subsequently showed that a cyclic carbonate, 6-methoxybenzoxazolinone (6-MBOA), extracted from green wheat of about 10 cm in height induced breeding in a natural winter population of *M. montanus.* It appears that the end of the breeding season is not only the result of a lack of 6-MBOA, but also the result of inhibitors of reproduction, such as cinnamic acids, which are relatively abundant in grasses after they have flowered, fruited, and become dehydrated, which usually occurs in late summer and fall. Under laboratory conditions, the presence of green feeds, such as wheat, grass, or alfalfa, in the diet is essential for breeding of *M. montanus* and *M. arvalis.* Photoperiod does have a modulating effect on reproduction in *M. agrestis* and *M.arvalis,* to the extent that 15-20 h light per day is optimal, although *M. arvalis* will reproduce, but at a lower rate, under a 5L:19D regime. In *M. arvalis,* testicular weights are less under 6L:18D than under 18L:6D, but some spermatogenesis occurs under the short photoperiod.

As in golden hamsters, the pineal exerts an inhibitory effect on reproduction of *M. agrestis* kept under short photoperiods. For example, destruction of the sympathetic nervous system by 6-hydorxydopamine, which blocks the transmission of information of photoperiod to the pineal, and pinealectomy each stimulated gonadal activity of both sexes kept under short photoperiods. Negus and Berger (1971) reported that the feeding of green foods reduced the pineal weight of *M. montanus* kept under 8L:16D, 12L:12D, and 16L:8D. It is not clear, however, what is cause and what is effect. It is conceivable that pineal weight decreases as a result of the higher GTH and steroid hormone concentrations in the animals fed greens, and it is also possible that the animals fed greens, and it is also possible that the feeding of greens inhibits pineal activity and thus permits gonadal activity even on short days. Presumably, on long days, the pineal has little inhibitory effect on reproduction of this species.

In laboratory strains of *M. argestis,* long photoperiods, 16L:18D, stimulated testicular and ovarian activity independent of the ambient temperature (6.5 ± 1.5^{0}C and 20.0 ± 2.0^{0}C). In wild-trapped *M. agrestis,* long photoperiods maintained testicular activity of sexually mature animals, but with 6L:18D at 20 ± 2^{0}C testicular weights were also maintained, which did not occur in the laboratory strain. Roth (1974), using red-back voles (*Clethrionomys gapperi athabascae*) caught in the wild, found no effect of photoperiod or temperature and no interaction between these environmental variables in a factorial experiment with 20L:4D, 4L:20D, 20^{0}C, and 5^{0}C as the variables. In all four environments, gonadal size increased to about the same extent.

Temperature

It is, of course, of interest of know to what extent ambient temperature determines the start or the end of the reproductive period or to what extent ambient temperature affects reproductive performance. As Wimsatt (1969) pointed out in his review, there appears to be a mutual antagonism between hibernation and reproduction. For example, when female thirteen-lined ground squirrels (*Citellus tridecemlineatus*) are maintained at room temperature without hibernation, reproductive development is inhibited compared to controls that are kept in cold and darkness for several months, during which time they hibernate. Such inhibition is not found in males. During the breeding season, exposure to low temperature does not induce either testicular regression or hibernation and, as long as the males are maintained at these low temperatures, testicular regression is prevented. In the spring, males and females may hibernate, but the tendency to do so decreases as gonadal activity increases. During this spring phase, gonadectomy of either males or females increases the tendency to hibernate; however, this effect is restricted to the spring.

Prolonging the winter environment in the spring by keeping the animals in a cold and dark environment delays the onset of reproduction of chipmunks (*Tamias striatus*). In this case the effects of temperature and photoperiod cannot be separated. The length of the gestation period of the pippistrelle bat (*Pippistrellus pippistrellus*) is extended by exposure to a temperature between 5⁰C and 14⁰C, and it is shortened by exposure to temperatures of 35⁰C.

Ambient temperature can affect mammalian reproduction of non-seasonally reproducing species. House mice can be bred for many generations at 3⁰C; however, comparison with mice bred at 21⁰C has shown that at 3⁰C there are more barren pairs, and fertile pairs rear fewer young, although litters are larger at birth and the young are heavier, so that the smaller number of young reared is the result of greater postnatal mortality. Mice that had been transferred to 21⁰C were more fertile than controls bred at 21⁰C for 10 generations.

High ambient temperature seem, in general, to have more deleterious effects on reproductive processes than low ambient temperatures. Effects of high ambient temperature on reproduction have been studied in some detail in sheep, when sheep were exposed during different stages of the reproductive cycle to high ambient temperatures (40⁰C), it was found that (1) the unfertilized egg was not damaged; (2) spermatozoa in ewes at the time of capacitation were capable of fertilization, but there was a high pre-implantation mortality; and (3) exposure when the fertilized egg was undergoing its first cell division resulted in post-implantation mortality.

Olfactory Signals

The mammalian species discussed so far in this chapter have seasonal breeding, and we have discussed the effect of photoperiod, photoperiod and temperature, and the availability of certain feeds as proximate causes for the timing of reproduction. In the species to be discussed in this next section, there may be no breeding season, and the environmental factors to be discussed are factors that are created with the population itself. Olfactory signals in the form of chemicals, of which the identity is not always known, may be excreted in the saliva, the urine, in the feces, or secreted from specialized skin glands. They announce the presence of conspecifics to each other, and they may also serve to announce the sex and sexual status of the emitter of the signal.

Priarie voles

Estrus in the pariarie vȯle (*Microtus ochrogaster*) is induced, and the induction occurs through the male urine. Under laboratory conditions, olfactory signals have several important effects on the reproductive physiology of the female.

Olfactory urinary signals emitted by the male affect female reproduction by advancing puberty, inducing estrus in sexually mature but reproductively inactive females, and interrupting pregnancy induced by another male. Advancement of puberty has been found when female prairie voles were exposed to an unfamiliar adult male or to a male of the same age but from a different litter. It remains to be determined whether this advancement of puberty is influenced by prepuberal exposure to males.

The induction of estrus is clearly an advantage for the male, since it allows him to copulate and thus transmit the alleles to the genes he carries to the next generation. This advantage is based on the assumption that induction of estrus also occurs in feral populations. In the laboratory, induction of estrus occurs only when male and female are unfamiliar with each other. For example, an adult male in a cage with his female partner and one or more of the litters of the pair does not induce estrus in his daughter, unless he is removed for 8 d or more and then returned to his cage. Upon return, he induces estrus in most of his daughters of 36 d of age and all his daughters of over 45 d of age. If a female in a pair has become anestrus, removal of the male and return after 8 days or more, brings her into estrus. Evidence that the signal is an olfactory one and that it is present in the urine is provided by different experiments. First, anosmic females will not come into estrus. Second, the placement of one drop of urine on the upper lip of reproductively inactive females causes a rapid increase in LH-RH in the olfactory bulb, an increase in serum LH concentration, and a decrease in nrepinephrine concentration in the posterior olfactory bulb. Once the female is in estrus, she will remain so for as long as a month if she does not copulate. After copulation, which induces ovulation, the presence of the male is required for at least 24 h for pregnancy to continue. If the male is removed within 24 h post coitus, only about 35 percent of the females remain pregnant; if the male is removed 5 d post coitus, 95 percent of the females remain pregnant. This effect of the presence of a male in maintaining pregnancy is probably not the result of olfactory signals, but the result of additional copulations that are necessary to maintain the CL of pregnancy. Finally interruption of pregnancy by a strange male occurs if the stud male is removed, then when a strange male is placed with the female, the pregnancy induced by the stud male is interrupted and the female returns to estrus. This interruption of pregnancy can occur as late as day 17 of pregnancy, with a normal grestation length of 21.4 d in this species.

Males not only emit olfactory signals, they also receive them from females and other males. Male prairie voles show a preference for estrous vaginal and estrous urine odors, a preference that is androgen-dependent. In a population of male prairie voles that are behavioural compatible, either the introduction of female estrous odors or making the males anosmic, results in increased threatening and increased aggression, suggesting that odors serve as important signals among males for recognition and avoidance of aggressive encounters.

House mice

In feral populations, which are usually of low density, reproduction of the house mouse (*Mus musculus*) is seasonal. The proximate causes, which time the breeding in these populations have not been determined, although availability of food to meet the caloric intake required for reproduction is the most likely single candidate; however, factors such as photoperiod and temperature may modulate the requirement for calories or may modulate the neuroendocrine axis. Commersal populations, closely associated with humans, usually have high density and usually show no seasonality. With commensal and laboratory populations of mice, olfactory signals have been found to be of primary importance in the regulation of reproduction. Figure illustrates many of the interactions that have been found, and we will discuss these in sequence.

Mouse populations typically live in a territory that is defended by a single male, who marks the boundaries of his territory by urinary markings. Within the territory live the male, ten or fewer breeding females, some of their offspring, and some subordinate males. Females and males use urinary markings to indicate their presence and to affect the reproduction of other conspecifices. The marking behaviour by the male is androgen-dependent and status-dependent; that is, subordinae males urine-mark infrequently and the territorial dominant male may make as often as 3,000 times per night. The female's urinary marking is not dependent upon her endocrine status, but is increased by the presence of an intact male but not by a castrated male. A male exposed to the urine of a female will show a increase in plasma LH and testosterone concentrations. The male habituates to the female cue and his response diminishes upon continued exposure to the same female or her urine.

The function of this elevation of LH and testosterone concentrations is not immediately obvious, since spermatogenesis is a continuous process, and under conditions in which the male is not exposed to female urinary odors. LH and testosterone are released episodically. Bronson (1979b) has proposed that the elevated testosterone secretion in response to female urine may increase the synthesis of the male's own urinary olfactory signals. Male urine has several effects on female mouse reproduction: (1) acceleration of puberty, (2) dissipation of the estrus-suppressing effect of grouping of female mice, (3) synchronization of estrous cycles, and (4) interruption of pregnancy induced by another male.

1. *Acceleration of puberty.* The induction of puberty in females at an earlier age than in females not receiving the olfactory male urine signal is called the Vandenberg effect after its discoverer. The puberty-accelerating effect is greatly enhanced if it is combined with tactile stimuli. Kirchhof-Glazier (1979), using mice pups nursed by prairie deer mice (*Peromyscus maniculatus bairdii*) mothers, showed that this effect was genetically determined and not acquired. Massey and Vandenbergh (1981) found that male urine collected from a feral population at its lowest density in June and at its highest density in January accelerated puberty in laboratory mice. The potency of the puberty-accelerating factor was similar in the samples collected in June and January, suggesting that population density did not affect the excretion of this factor.

The hormonal changes in the female prepubertal mice in response to the puberty-accelerating urine consist of an increase in plasma LH concentration followed by an increase in plasma estradiol concentration. On day 3, the estradiol concentration returns to its base level, and, provided that the mechanism for the stimulatory effect to release LH has matured, is followed one day later by a preovulatory FSH and LH release followed by ovulation. The chemical that causes the acceleration of puberty is probably a protein. It is present in bladder urine and is excreted with the urine, and its production is androgen-dependent.

2. *Dissipation of the estrus-suppressing effect of grouping of female mice without males.* Female mice grouped together show a high incidence of pseudopregnancies in the absence of mating. The presence of a male or as little as 0.1 ml of male urine introduced

into the cage with grouped mice results in resumption of normal 4- and 5-d cycles.

3. *Synchronization of estrus.* In a population of mice with 4-d cycles, one expects 25 percent of the population to be in estrus on any given day, as is generally found unless the van der Lee-Boot effect interferes. If a male or male urine is introduced in a population of regularly cyclic female mice, a peak of estrous females occurs on day 3 after introduction of the male. This is called the Whitten effect, and it is not restricted to mice, but also occurs in sheep and goats. The material excreted in the urine that induces the Whitten effect in mice is probably lipoidal and is present in the urine in the bladder. The excretion of the chemical is androgen-dependent but does not involve the secondary sex organs or their secretions. The effect of the male olfactory signals in synchronization of the estrous cycles of mice seems to be the result of a depression of progresterone secretion within 36 h after exposure to the male. This is followed by an increase in plasma estradiol and by a preovulatory peak in plasma LH concentration.

4. *Interruption of pregnancy by a strange male.* If the male of a mated pair of mice is replaced by another male, i.e., a strange male within the first five days post coitus, pregnancy is interrupted. This pregnancy-blocking effect or Bruce effect is followed by the induction of estrus. The greater the genetic difference between the stud male and the strange male, the higher the incidence of blocked pregnancies. The effect of the strange male can be simulated by exposure of the females to the urine of a strange male. The production of the chemical(s), a peptide, that cause(s) the Bruce effect is androgen-dependent, but the site of its production has not been ascertained.

The endocrinological changes in the pregnant female in response to a strange male are probably interruption of prolactin secretion and release of LH, which are necessary to induce the new estrous cycle and ovulation. Chapman et. al. (1970) suggest that increased LH secretion proceeding a decrease in prolactin secretion interrupts progesterone secretion by the CL of pregnancy and thus causes the pregnancy block. This hypothesis is based on changes in AP content, not on changes in plasma concentrations of LH and prolactin; therefore, it needs further substantation.

The Bruce effect has also been observed in *Microtus ochrogaster*, *M. agrestis*, *M. pennsylvanicus*, and *Peromyscus maniculatus bairdii*. The Bruce effect probably is rarely, if ever, encountered in feral populations and may be a result of artificial selection.

Suppression of estrus by females

Female mice emit urinary signals that suppress estrus in prepubertal females as they approach puberty and in adult female mice when adults are grouped together. Exposure of young female mice to urine from adult or juvenile grouped females for 7 d during the first two weeks after weaning delayed the onset of puberty by 3-5 d in comparison with controls that were not exposed to urine. Tactile stimuli by the grouped mice had no effect on the urinary cue effect, in contrast to the enhancement by tactile stimuli of the puberty-accelerating effect of male urine. Massey and Vandenbergh (1980) have shown that urine from grouped female laboratory house mice was effective in delaying puberty in a feral mouse population.

When female mice are groped without a male or without exposure to odors form male urine, estrous cycles become irregular; there may be a high incidence of pseudopregnancies in the absence of any copulations; or there is a high incidence of anestrus but without any CL formation. The occurence of either of these two forms of interrupted estrous cycles seems to depend on the number of mice housed together, with about 4 mice there is a high incidence of pseudopregnancies; with 30 mice there is a high incidence of anestrus. As mentioned in the discussion of olfactory cues emitted by male mice, exposure of such grouped mice to male mice, exposure of such grouped mice to male urine eliminates the inhibitory effect of female olfactory signals on the estrous cycle.

The pathway by which odors affect the neuroendocrine unit probably includes the vomeronasal organ→the vomeronasal nerve→the accessory olfactory bulb→the medial and posteromedial amygdaloid nuclei→the stria terminalis→the medial preoptic area, ventromedial nucleus, nucleus tuberis lateralis, and premammliary nucleus. There is persuasive evidence that sectioning of the vomernasal nerve of hamsters interrupts copulatory behaviour, whereas destruction of the nasal mucosa does not. The importance of this accessory olfactory system may explain the significance of the Flehmen behaviour in bulls, stallions, rams, and other mammals. This Flehmen is the moving up of the head and the curling of the upper lip. This behaviour may allow the chemical signals easier access to the vomeronasal organ.

Wild mice are nocturnal, and in spite of many generations of laboratory breeding, laboratory mice showed a preovulatory LH peak that was 2.4 times greater under DD than under 14L:10D. A similar peak was observed in ovariectomized estrogen-treated mice in which

LH release was induced by progesterone. The number of oocytes ovulated, however, was not affected by the light treatment, and a tenfold increase in light intensity did not affect the peak LH concentration. However, wild mice caught and exposed to 14L:10D at 1000 lx showed a depression in body weight of males, and an over 50 percent reduction in litter size, whereas a strain of laboratory mice did not show these effects. Under DD, both stocks reproduced normally.

Rats

Two types of populations of wild brown rats, from which most strains of laboratory albino rats are derived, exist; these are the feral populations that exist independent of human activities and the populations closely associated with human activities, such as city rat populations. Much of the present knowledge about this species is based on the laboratory strains, which may be quite different from the wild populations. For example, McClintock and Adler (1979) found that the estrous cycle of wild rats caught in a city and then studied in the laboratory had an estrous period of one or two weeks instead of the 4 or 5 days found in laboratory rats.

Steniger (1950), in a study of a feral rate population in North Germany, found no pregnant rats between October and March. He attributed this seasonal reproduction to the influence of temperature, but provided on data to substantiate this claim. In a city population, Davis (1951) found evidence for some seasonal fluctuation in the incidence of pregnancy, but pregnant rats were found throughout the year. Calhoun (1962), in an extensive study of rats caught in a city and studied in an outside pen with a natural environment, found clear evidence of seasonal reproduction by examination of the nipples of females and by diagnosis of conceptions. In this population, no pregnancies occurred between September and January. Temperature was not the major factor controlling reproduction, because rats conceived in February when the mean minimum nightly temperature, -3^0C was below the 15^0C of September. Calhoun (1962) suggested that photoperiod might be a controlling factor. Food was always available in abundance, thus an indirect effect of season through an effect on food availability can be eliminated as one of the proximate causes for the seasonal reproduction. At present, too little information is available to state what the proximate causes of the seasonal reproduction of wild rats are.

The estrous cycle of laboratory rats is affected by photoperiod; under 12L:12D rats showed exclusively 4-d cycles, whereas under

14L:10D and 16L:8D, there was a high incidence of 5-d cycles. In most strains of laboratory rats, exposure to LL induces a permanent vaginal estrus syndrome (PVE), which is characterized by a failure of ovulations to occurs, cystic ovaries, and persistent estrus shown in vaginal smears. After a few days of exposure to LL, the preovulatory LH surge disappears. Recent investigations have shown that in rats under LL, taking smear at 24-h intervals instead of a random times each day delayed the onset of the PVE. This suggests that the daily light-dark periods may act as synchronizers and that LL causes PVE partly by the removal of such a signal and partly by an inherent effect on the neuroendocrine unit.

For LL to have the effect of inducing PVE, the light needs to be bright. Weber and Adler (1979) subjected rats to either LL, LL or a light regime providing a contrast in light intensities, 12L:12L. On the 12L:12L regime, fewer animals came into constant estrus than under either of the other two LL regimes, again pointing to light functioning as a cue instead of as a stimulator or inhibitor.

During solicitation and mounting, males emit ultrasounds of 50 KHz. The number of such calls are higher before successful than before unsuccessful mating and are higher before mounts with intromissions than before mounts without intromissions. During the postejaculatory refractory period, the male rat emits ultrasounds of 22 KHz, which are correlated with a sleeplike electroencephalogram pattern. Barfield *et. al.* (1979) suggest that the 22- KHz vocalizations may serve to maintain contact between partners, at the same time preventing too early recopulation. Adler and Anisko (1979) submit that the 22-KHz ultrasounds are uttered when the rat is in a submissive or helpless state. The emission of postejaculatory ultrasounds by male rats does not appear to depend on the presence of testicular hormones, but estradiol dipropinoate has an inhibitory effect.

The production of ultrasounds my involve the central gray area. The pathway by which the ultrasound affect the neuroendocrine unit involved in presexual, sexual, and postsexual behaviour needs to be elucidated.

Olfactory communication occur among rats as they do in hamsters and mice, but the effect of female olfactory signals on other females are different in rats than in mice. In grouped rats and in rats housed singly but exposed to odors from grouped rats, the variation in the length of the estrous cycle was shorter than in isolated rats housed single rats exposed to odors from grouped rats became synchronized, i.e., a majority of females came into estrus on the same day. Some ingenous experiments have provided evidence that the odors from rats with the LL-induced PVE syndrome impart the acyclicity on other rats downwind, but not in rats upwind with respect to the PVE rats. Odors from proestrous rats desynchronized the synchronous estrous cycles of rats housed together. Odors from metestrous rats had a synchronizing effect on rats in different stages of the cycle. It thus appears that proestrous rats are the leaders in synchronizing the estrous cycle in grouped rats. The proestrous rat first gives the signals that tend to synchronize the rats and then when the proestrus rat goes into estrous, she emits an olfactory signal that brings the group into synchrony.

In addition to the male-female olfactory communication, there is olfactory communication between the lactating mother and her offspring. The extensive review and discussion by Leon (1978) should be consulted for these interrelationship in rats. In brief, it appears that there are: (1) maternal odors that inhibit gross motor activity between age 2-12 d, (2) olfactory cues necessary for orientation of the pup towards the nipple, and (3) home odor cues that allow the pup to orient toward the nest. This cue may be imparted by the mother to the bedding. That the mother emits olfactory cues attractive to the pups has been shown by a number of experiments. The maternal odor of lactating females is derived from her excreta, and is attractive to pups between the ages of 12 d and 27 d. It is apparently present in the caecotrophe (a semisolid, light-coloured material formed in the saecum and excreted through the anus). Lactating rats excrete more caecotrophe than nonlactating females. The difference between the attractiveness of exceta from virgins and lactating rats may be the result of the greater food consumption by the lactating females and the greater amount of caecotrophe excreted in conjunction with this. The amount of food ingested in related to the number and age of pups being nursed. The emission of the odor is correlated with high prolactin secretion, which, in turn, increases food intake.

The odor is apparently synthesized by caecal microorganisms. The weakening of the pheromonal bond is probably the result of ingestion of maternal fecal material and the subsequent production of their own caecotrophe and their own odor, which resembles that of their mother. At the same time, this odor production by the mother diminishes and the attraction of the young to the feces of the mother is transferred to their own feces. As rats normally eat their own feces, such a transfer makes evolutionary sense.

The stimulus provided to a male rat by an estrous female may well consist of a complex mixture of auditory, olfactory, and tactile signals. Even when the exact nature of the stimulus is not known, it is possible to obtain the hormonal response in a male rat by classical conditioning of the male. Exposure of sexually naive males first to a neural stimulus, which did not result in LH release and testosterone secretion, and then to an estrous female once daily for 14 trails, resulted in LH release and testosterone secretion after exposure to the neural stimulus alone.

Rhesus monkeys

Captive rhesus monkeys (*Macaca mulatta*) in an outdoor environment show a definite seasonal pattern of reproduction. The onset of mating activity occurs in late summer to early fall and continues for 4 to 5 months. Individual females are receptive for 5-11 d during the follicular and periovulatory part of the cycle. Rhesus monkeys kept in a laboratory environment show no seasonal pattern of reproduction and females are receptive every day of the menstrual cycle, although there is a peak of copulatory activity around the time of ovulation. The proximate cause for the seasonal reproduction of free-ranging rhesus monkeys has not been determined. Within laboratory populations of rhesus monkeys, much research has been done on the role of olfactory cues emitted by the female in the initiation of mating behaviour by the male. Goldfoot (1981) has reviewed both the evidence and the interpretation of the evidence. It appears that for sexually experienced males, odors from midcycle products of the vagina make a female more attractive; however, such olfactory cues appear neither necessary nor, by themselves, sufficient to induce matings.

Nutrition

Certain foods can act as proximate factors in timing of the breeding season, as was discussed for *Microtus montanus*, *M. agrestis*, and *M. arvalis*. Nutrition can also affect reproduction without serving as a proximate cause. In general, little is known about the nutritional requirements of wild populations, whereas the nutritional requirements of some laboratory and farm animals are known in considerable detail. It lies outside the scope of this book to review the effects of specific nutrients on reproduction, but in most cases the deficiency of a nutrient leads to inanition, so that it becomes difficult to assess whether a specific nutrient deficiency affects reproduction. Wild populations may occasionally face deficiencies of water, calories, and proteins, and we will discuss there briefly. Water deprivation, even in the gerbil

(*Meriones unguiculatus*), a desert animal, leads to servere depression of rêproduction, without affecting ovarian or testicular weights. Gerbils do not need water for survival, because they an use the water present in food and metabolic water; for reproduction, however, water is a requirement. Deficiencies of caloric intake may range from intermittent deficiencies to total starvation. On theoretical grounds, one would expect caloric deficiency, especially for a protracted period, to have the severest repercussions on females in which the oocyte contains large amounts of yolk, such as the elasmobranchs, reptiles, and birds, whereas the effect on males might be expected not to be as severe as the effect on the females. In species that incorporate little yolk in the oocyte, one might expect that caloric deficiency would not affect ovulation and fertilization, but might result in abnormal embryonic development or rejection and resorption of the embryo.

Experiments with mice have shown that fasting of grouped males and females for as little as 18-30 h reduced fertility without affecting estrous behaviour; however, starvation for 36 h diminished estrous behaviour. That the effects were the result of a deficient caloric intake was shown by the fact that feeding a 75 percent glucose solution prevented the effects of starvation.

Withdrawal of food from mice for the 48-h period from day 3 to 5 after copulatory plugs were found, resulted in failure to give birth to young. Withdrawal of food at other times between day 1 and day 7 after copulatory plugs were found had less effect on the birth rate. Apparently, at the time of implantation, the embryo-uterine unit is particularly sensitive to caloric deficiency. This effect is mediated through the CL, since injections of either hCG or progestrone can prevent this effect of inanition of the mother.

Many experimental procedures have been used to study the effects of malnutrition (e.g., feeding of protein-free diets, total starvation, vitamin-deficient diets) on reproduction of female rats. In many of these cases, the embryos are resorbed or aborted, and these effects can be prevented by progesterone administration. Srebnik et. al. (1978) have presented evidence that malnutrition depresses the activity of the neuroendocrine unit but that it is the hypothalamus that is supersensitive in malnourished rats. They speculate that steriodogenesis is affected first and that failure of gonadotrophin secretion is an effect of the failure of steroidogenesis instead of its cause.

Starvation of laying chickens results in atresia of the large follicles which, of course, results in cessation of egg production. Egg production can, however, be maintained for about 10 days by the injection of either mammalian FSH and LH or by chicken AP extracts. The eggs become smaller as starvation progresses in such gonadotrophin-injected hens until the reserves that can be mobilized are exhausted.

Deficiency of calcium in the diet of laying hens leads to inanition and to a severe reduction in egg production and an increase in the proportion of soft-shelled eggs. Injection of chicken hypothalamic extracts does not increase the number of eggs laid during a 21-day period of calcium deficiency, but it does increase the ratio of hard-shelled to soft-shelled eggs compared to calcium-deficient hens injected with saline or mammalian LH, mammalian FSH, or chicken pituitary extracts. It would be of interest to follow the concentrations of steriod and gonadotrophic hormones during the feeding of such a calcium-deficient diet and to determine whether the stimulatory effect of progesterone on LH secretion and vice versa is maintained in such hens.

Several food sources have been identified as inhibitors of reproduction, skunk cabbage and the bush *Salsola tuberulata*. Both caused the failure of labor, because the damaged pituitary development of the fetus. In Australia, it has been found that sheep grazing on subterranean clover, *Trifolium subterraneum*, become sterile largely because of estrogenic compounds present in this clover.

It is beyond the scope of this book to review or discusss the effects of pesticides, fungicides and herbicides or of drugs, alcohol, caffeine, nicotine, etc. on the reproduction of laboratory animals, wild animals, or humans.

Population Cycles

The number of animals in populations of small rodents, such as mice, voles, and lemmings, fluctuates with a periodicity of about three to four years. One explanation of this periodicity is that in high-density populations, of this periodicity is that in high-density populations, more agonistic encounters occur that lead to an increase in secretion of ACTH corticosterone. As the population density increases, high concentrations of corticosterone become more or less permanent and start to have deleterious effects on reproductive performance and survival of the young. Laboratory experiments with mice tend to support this hypothesis, but whether it explains fluctuations in feral populations is still a matter of controversy. Myers and Krebs (1974) suggest that, within populations of rodents showing these fluctuations in density, there may be two genotype, one intolerant of high population density but reproductively superior, and another genotype with a low rate of reproduction but adapted to high density. They suggest that as the population density increases, the individuals intolerant of crowding, migrate, thus decreasing the population density.

The relationship between stress by agonistic encounters, reproduction, and population density is illustrated in a unique manner in the genus *Antechinus* of dasyurid marsupials. In *A. stuarti*, *A. swainsoni*, *A. minimus*, *A. flavipes*, *A. bellus*, in *A. bilarini*, the males die soon after mating, in *A. stuarti* as early as three weeks post coitus. In this species, mortality is associated with several symptoms considered as reliable indicators of stress, such as high plasma corticosteroid concentrations, high liver glycogen content, high adrenal weight, and a wide fasciculate zone. Not only does plasma corticosteroid concentration in males rise from 2.9 ng/mal prior to mating to 5.2 ng/ml after mating, but there is also a decrease in the corticosteroid-binding capacity of the serum, thus making more free corticosteroid available. Females, in contrast to males, are capable of reproduction during two or three consecutive years.

Injections of cortisol acetate into captive males of *A. stuarti*, caused mortality in 9 of 27 animals, but no death among 11 controls. Mortality of males captured prior to mating and kept isolated in captivity survived beyond the normal life expectancy, but males captured during the last week prior to the expected time of death did not survive in isolation. The apparent stress which causes the syndrome, may consist of agonistic encounters and copulations. The phenomenon of once-in-a-life-time reproduction, which has been described for lamprcys and salmon previously in this book, has been called *semelparity*. Lee et. al. (1977) speculate that the high male mortality in brown lemmings (*Lemmus trimucronatus*) and in *Microtus pennsylvanicus* during highly synchronized spring mating may be similar to the semelparity of *A. stuarit*.

One researcher noted that when the tree shrew *Tupaia belangeri* is disturbed, the sympathetic nervous system is stimulated; which results, among other events, in activation of the arrectores pilorum muscles which raise the hairs on the tail. By expressing the time that the animal has a bushy tail as a percentage of the time that the animal was observed, an indicator for the amount of disturbance was obtained. Most of the bushy tail resulted form confrontation with a dominant male, and von Holst (1969) found a strong relationship between the percentage of time in animal had a bushy tail and reproductive disturbances. When the percentage of time of bushy tail was between 20 and 40 percent, female tree shrews showed male in stead of female sexual behaviour and a high frequency of eating of their young, presumably because they had failed to mark them with their sternal scent gland. When the percentage of time that the animal had a bushy tail was 90 percent or more, females were sterile and had small ovaries without mature follicles. In lactating females, when this percentage was above 20, milk secretion descreased. In tree shrew males, the testes were in the scrotum but could be retracted into the abdomen when the percentage of time that the animal had a bushy tail was between 50 to 70 percent; above 70 percent the testes remained in the abdomen, and the germinal epithelium degenerated; whereas below 50 percent the testes remained in the scrotum and were not retracted into the abdomen. These data von Holst (1969) interpreted as indicating that confrontations with a dominant male stimulate the sympathetic nervous system, resulting in raising of the hairs on the tail and also affecting the hypothalamic neuroendocrine unit, so that the secretion of FSH, LH, TSH, and growth hormone are decreased, but ACTH secretion is increased. This unusual method of evaluating stress should be used in conjunction with determinations of corticosteroids, gonadal, and the pituitary hormones mentioned in order to verify von Holst's interpretation (1969).

Index